THE INTELLIGENT BUILDING SOURCEBOOK

THE INTELLIGENT BUILDING SOURCEBOOK

Johnson Controls, Inc.
John A. Bernaden and
Richard E. Neubauer, Editors

Library of Congress Cataloging-in-Publication Data

The Intelligent building sourcebook.

Includes index.
1. Office buildings -- Automation. 2. Office buildings -- Environmental engineering. 3. Office buildings -- Communication systems.
I. Bernaden, John A.   II. Neubauer, Richard E.
III. Johnson Controls, Inc. IV. Title: Intelligent building sourcebook.
TH6012.I57 1988   696   85-45877
ISBN 0-88173-019-X

The Intelligent Building Sourcebook.

Published by The Fairmont Press, Inc.
700 Indian Trail
Lilburn, GA 30247

ISBN 0-88173-019-X   FP

ISBN 0-13-468935-6   PH

While every effort is made to provide dependable information, the publisher, authors, and editors cannot be held responsible for any errors or omissions.

Printed in the United States of America

Distributed by Prentice Hall
A division of Simon & Schuster
Englewood Cliffs, NJ 07632

Prentice-Hall International (UK) Limited, London
Prentice-Hall of Australia Pty. Limited, Sydney
Prentice-Hall Canada Inc., Toronto
Prentice-Hall Hispanoamericana, S.A., Mexico
Prentice-Hall of India Private Limited, New Delhi
Prentice-Hall of Japan, Inc., Tokyo
Simon & Schuster Asia Pte. Ltd., Singapore
Editora Prentice-Hall do Brasil, Ltda., Rio de Janeiro

ACKNOWLEDGEMENTS

A book like this <u>Sourcebook</u> is always the result of the efforts of many persistent and consciencious people working toward common objectives. Sharing knowledge about the many facets of building intelligence has motivated authors, review groups, editors, secretaries, public relations personnel, etc., all of whom deserve our gratitude and appreciation. To all of these co-authors we want to say "Well done!"

We would like to thank Mary Hollrith, Cynthia Newton and George Huhnke for their enthusiasm in completing this book. They have spent many hours working with all of the book's contributors.

Finally, appreciation is due also to Joe Lewis, Dave Bigler, Ron Caffrey and Dick Wilson, who have encouraged this book's necessity and effort.

John A. Bernaden and
Richard E. Neubauer, Editors

FOREWORD

The intelligent building concept is a dramatic growth market in the United States and it is spreading throughout the world. Building structure, comfort control, energy management, fire and security systems, advanced electronic systems and communications networks, shared tenant services -- all are involved.

Full use of available technology will increase the cost of construction by $3 to $10 per square foot or more, which is a 5 to 10 percent increase in overall construction cost. Some builders, developers, regulatory agencies, construction trade unions . . . some institutions of all kinds will resist changes for various selfish interests, but tenants are becoming aware of the benefits of intelligent buildings. They believe these benefits are tied to individuals' productivity. And the cost of the individual worker's salary and benefits far exceeds the cost of his portion of the building.

The economics, as well as the emotions of the business, will prevail and within five to ten years intelligent buildings will replace conventional buildings--just as elevators and air conditioning became standards in their own ten-year periods in the past.

The universal truth about real estate is indeed being rewritten. The formula for a successful real estate development was simple. It was "location-location-location." Today, it is becoming "location-location-<u>utility</u>" (by utility I mean effectiveness for the occupant, i.e., <u>productivity</u>).

Like many new ideas, understanding the subject is not universal. This book can contribute greatly toward separating hype from reality; toward understanding the true nature of the systems and services available. The concept foretells increased prosperity for both builders and occupants--a win-win situation.

When one of these situations exists it deserves broad attention and effort.

For Johnson Controls, and for the Intelligent Buildings Institute, I commend and thank the authors.

Ronald J. Caffrey
Intelligent Buildings Institute Chairman
Vice President, Marketing
Johnson Controls, Inc.

CONTENTS

Section IV - FUNCTIONAL INTEGRATION

Section V - SHARED TENANT SERVICES

Section VI - CASE STUDIES

Section VII - DEVELOPING TOMORROW'S BUILDING TODAY

SECTION I

INTELLIGENT BUILDING AUTOMATION

CHAPTER 1

The Intelligent Building: An Overview

Richard E. Neubauer
Johnson Controls, Inc.

Bursting upon the scene in recent years, the "Intelligent Building" has captured the imagination and excited the interest of a wide public ranging from building owners, developers, managers and tenants to architects, engineers, realtors and consultants. However, although there is a new generation of intelligent buildings currently unfolding, the concept of building intelligence is not new.

BUILDING INTELLIGENCE IS A CONTINUUM

Building intelligence consists of a continuum of capabilities provided by a variety of information services or systems. Based on this definition, buildings with some type or degree of intelligence have existed for many years. What is new today is that the development of building intelligence is taking a giant leap forward as a result of the constantly increasing capabilities and decreasing costs of microelectronic technology, an increasingly more favorable regulatory environment, a progression of standards facilitating system integration, and a growing awareness of the benefits of intelligent building operations.

To put this development into perspective, it is helpful to remember the basic reason why buildings are constructed. Buildings are constructed to meet human needs.

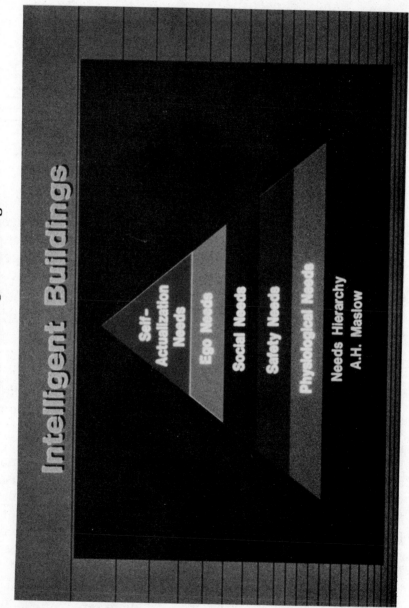

Fig. 1-1 Intelligent Buildings

The noted psychologist, Abraham Maslow, developed a hierarchy of human needs. Designed in the form of a triangle, this hierarchy begins with physiological needs at its base. It then ascends through safety, social and ego needs levels to self-actualization needs at its apex. Maslow made the point that people progress on a first-things-first basis from the base to the apex of the triangle, satisfying the needs of one level before proceeding up the triangle to the next level. (See Figure 1-1)

This hierarchy of human needs applies to buildings, and in particular intelligent buildings, as well as to every other area of life. People initially use buildings to meet their physiological needs in terms of heating, air conditioning, ventilation, light and water. Next, buildings are used to meet safety needs from the standpoint of security and fire protection.

Building intelligence then appears in the form of information systems designed to better meet physiological and safety needs by automatically monitoring and managing energy consumption and security and fire protection. Telecommunications systems are implemented to meet needs for social or verbal communications. Office automation systems are developed to satisfy data and word processing needs. Other systems such as videoconferencing and satellite communications may also be added to further respond to the ever-rising needs of building occupants.

Ultimately, the intelligent building becomes a powerful, dynamic tool which can be used to create the personal, environmental and technological conditions necessary for building occupants to maximize their individual capabilities, productivity and satisfaction.

In other words, the role of the intelligent building is to be a servant of human needs, not an example of technology for technology's sake.

TYPES OF BUILDING INTELLIGENCE

Building intelligence can consist of a number of different information systems. These systems can vary greatly in terms of their capabilities, including the degree to which they are integrated with each other.

Most buildings today contain one or more of these information systems and are intelligent to some degree. So it is not a question of intelligent buildings on the one hand versus "dumb" buildings on the other. Rather, all buildings exist on some point of a continuum of capabilities ranging from the least to the most intelligent.

Building Automation System

Building intelligence starts with a monitoring and control information service called a building automation system (BAS). Operating with temperature, pressure, humidity, time, and other data, this system is designed to automatically maximize the comfort of the occupants of a building while minimizing the energy used by the building's heating, ventilating, and air conditioning (HVAC) equipment. The system can also be used to achieve economical, efficient lighting control. Other system functions may include automated security and fire protection, depending on what type of integration of these functions is permitted under local codes. In addition, computerized building maintenance management can be included as an integral part of BAS operations.

Most buildings today have the beginnings of a BAS in the form of thermostats, time clocks, smoke detectors, door alarms, and other devices. But these devices are very limited in scope compared to the total environmental control, fire and security management, and other functions provided by a BAS with distributed processing and multiple-zone capabilities. Implementation of a building automation system is therefore an essential first step in the development of an intelligent building.

Telecommunications System

Another information service making up building intelligence is a telecommunications system. Most buildings today are equipped with analog PBX systems that are used for internal and external voice telecommunications. Analog PBX systems can also be operated with modems to make possible data as well as voice telecommunications. In addition, digital PBX systems are becoming increasingly available today. Providing the highest degree of sophistication, digital PBX systems make it possible to process voice and data telecommunications more efficiently.

Office Automation System

A third information service provided by an intelligent building is an office automation system. Most offices today have office automation in the form of data and word processing equipment. However, data and word processing operations often take place independently with little or no integration between the two functions. This results in considerable inefficiency. The preferred office automation system integrates data and word processing operations, making it possible to achieve full data integration capabilities.

Other Systems

Building intelligence primarily consists of information services provided by building automation, telecommunications and office automation systems. However, other systems are now gaining increasing acceptance in intelligent buildings.

One of these is the video communications system which makes possible visual teleconferencing. Another is the rooftop microwave or satellite system which facilitates bypass communications with other locations.

Fig. 1-2 PBX/BAS Integration

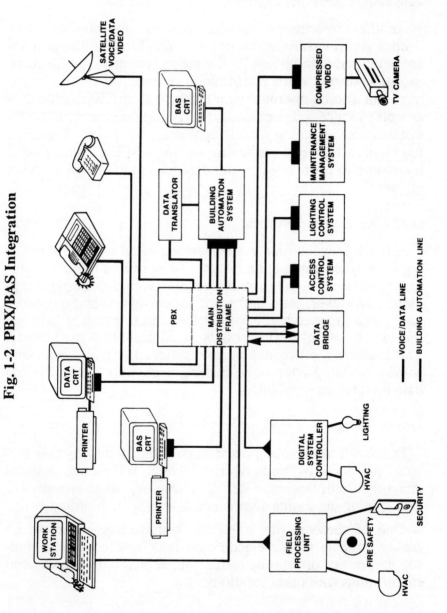

Conclusion

These systems enable buildings to handle information in all of its many forms -- data, text, speech and pictures -- both internally and externally. Depending on the degree of sophistication of each of these systems capabilities, this can result in a significant level of building intelligence.

However, one additional step is required to achieve a truly intelligent building. This step consists of integrating these systems so that they can fully interact with each other. Achievement of this goal requires that the systems first be structurally interfaced.

BUILDING INTELLIGENCE REQUIRES STRUCTURAL INTERFACING

Structural interfacing involves physically connecting two or more systems. This interfacing is necessary because the systems normally are separately wired. A BAS, for example, usually operates with a twisted-pair network in a multidrop configuration. A telecommunications system usually uses a twisted-pair network in a star configuration. An office automation system usually operates with conventional coaxial cable using a baseband or broadband local area network. A video communications system usually uses broadband coaxial cable. In addition, fiber optics cable is increasingly being used as a communications media in some of these systems.

Ways To Achieve Physical Connectivity

Physical connectivity can be achieved between systems in several ways. One way, for example, is to install gateways to link systems together. Another method is to operate several systems on the same wiring network. A BAS and a telecommunications system, for instance, can share the same twisted-pair wiring with conversion of the BAS configuration from a multidrop to a star design. Several systems can also be operated as part of a baseband local area network. (See Figure 1-2).

A more general approach is to use broadband coaxial cable as univeral wiring or, in effect, an "information highway." This is practical in cases where a wide range of systems, including video, is required to share a common communications network. Fiber optics cable can also serve as a high bandwidth, multiplexed digital communications media in a network, particularly in cases where immunity from electromagnetic interference and a high level of data reliability are required.

Interfacing Technology Is Available

The technology for developing structural interfacing is available today. It can be incorporated as an integral part of any new building. It can also be specified when space is leased in new buildings, many of which are now being equipped with varying degrees of intelligence. In addition, it can be implemented in existing buildings. But, since it is often difficult to establish economical feasibility in existing buildings, retrofits must be very closely analyzed to ensure that benefits outweigh costs.

INTELLIGENCE MAY INVOLVE FUNCTIONAL INTEGRATION

While integration of functions is not a requirement for intelligent buildings today, the capability to achieve total, functional integration of these systems is an ongoing process. Initially, development focused on integrating building automation and telecommunications or PBX systems.

Integration of these two systems makes it possible to use either special-feature telephones or integrated voice/data terminals to not only make phone calls but also access a building automation system. As a result, building occupants are able to "call up" the building automation system and enter commands to selectively turn HVAC equipment and lights on or off. In other cases, security personnel can call in checks while making rounds or alarms can be activated. Other possible applications based on the integration of building automation and telecommunications systems will almost surely appear.

Office Automation Integration Is Next

Work is proceeding on the integration of building automation and telecommunication systems with office automation systems via baseband, broadband and fiber optics networks. In addition, video communications and satellite systems will undergo integration in the future as the need arises.

Overall, the objective of this systems integration is to synergistically increase functionality while reducing costs. Based on meeting real human needs with cost-effective solutions dictated by economic and functional benefits rather than technological hype, achievement of this goal will enable the intelligent building to assume its ultimate role as a dynamic, adaptive and responsive tool of information systems management.

BENEFITS OF THE INTELLIGENT BUILDING

The economic and functional benefits of the intelligent building include longer building life, lower installation and life cycle costs, reduced energy consumption, higher space efficiency and greater worker productivity.

Flexibility is enhanced because the intelligent building uses reprogrammable, functionally-distributed systems that can be economically upgraded to meet changing requirements, minimizing the need for costly structural modifications while simplifying ongoing cable and wire network management. The intelligent building is also flexible in terms of the ability to not only tie together different systems but quickly, easily and economically locate workstations practically anywhere. In addition, the capabilities of the intelligent building can be flexibly expanded to accommodate future growth.

Increased Worker Productivity

Worker productivity is increased because all systems work together on an integrated basis, making it possible to improve efficiency while holding down personnel costs. At the same time, more efficient building operations are achieved at less cost because no more energy is used than necessary to provide adequate comfort levels in all building areas.

Summing up, the intelligent building provides better overall control of every facet of its operations. In the future, even more advanced controls and capabilities will make it possible to perform many additional and more sophisticated functions.

Advancing technology, for example, is currently creating an entirely new generation of sensors and control devices combining people-sensing with micro-environmental control. Other technological approaches will make it possible to anticipate rather than merely respond to changes in weather conditions and building occupancy. In addition, future development of the Integrated Services Digital Network (ISDN) will usher in a new, more cost-effective way of implementing a single, universal communications medium.

Owner/Developers Benefit As Well As Occupants

These benefits are provided to the occupants of an intelligent building regardless of whether the building is owner-occupied or tenant-occupied. In either case, employees or tenants obtain maximum benefits from working in an intelligent building.

In addition, owner/developers who construct commercial buildings and lease them to tenants benefit from the advanced information system capabilities provided by an intelligent building. Offered as shared tenant services, these capabilities serve as selling points which make it easier to lease intelligent building space.

INTELLIGENT BUILDING DEVELOPMENT

The intelligent building exists on a broad spectrum of capabilities. The degree of intelligence developed in any given building is a management decision.

An intelligent building may be developed by an owner for the use of employees. Or it may be constructed by an owner/developer for leasing to tenants. The degree of intelligence incorporated in owner or tenant-occupied buildings may vary depending on a wide variety of factors such as available technology, competitive considerations and the real estate market for commerical office building space.

System Integration Varies

An owner-occupied building may provide greater integration of systems than a tenant-occupied building in which the owner/developer is offering shared tenant services. The owner who constructs an intelligent building for the use of employees may be able to make a larger initial investment to achieve greater integration because of a longer timeframe in which to justify the investment. By contrast, the owner/developer who plans to lease intelligent building space to tenants may not be able to justify as large an investment because of a shorter timeframe created by the need to conform with the current market for commerical office space.

Many Types Of Intelligent Buildings

Intelligent buildings are primarily being developed today in the form of owner and tenant-occupied office buildings. However, intelligence can and is being developed in many other types of buildings such as manufacturing plants, hospitals, hotels, schools and government facilities. In the future, no building will be immune to this trend.

Conclusion

All of the players involved in the building industry -- owners, developers, managers, architects, engineers, realtors and consultants -- must constantly keep abreast of ongoing developments in intelligent buildings. This is because of the rapidity with which changes are taking place in this field.

It is also important to recognize that no one company has a "lock" on all of the integrated information systems used in an intelligent building. There are forces at work such as the increasing number of suppliers in the field and the unique requirements of individual projects which dictate that multiple vendors will be involved in the development of intelligent buildings not only today but in the foreseeable future. A multi-vendor approach therefore should be followed to achieve maximum flexibility in intelligent building development while ensuring that the best possible systems are used.

Finally, owner/developers of intelligent buildings should be aware of the increasing role now being played by systems integrators. These are companies that specialize in selecting, interfacing and integrating intelligent building systems. System integrators make it possible to economically and efficiently use a multi-vendor approach in the development of a leading-edge, state-of-the-art intelligent building.

Chapter 2

The Integrated Building Automation System

Terry R. Weaver, P. E.
Johnson Controls, Inc.

Integration may well be the most overused, misused, and maligned buzzword in the Building Automation marketplace in the late 1980's. In its most broadly defined sense it sometimes means as little as utilizing a single proposal binder to convey cut sheets of functionally and physically discrete systems. This end of the spectrum can be described as "Marketing Integration" or "Brochureware Integration." While there may, indeed, be some purchasing or performance leverage gained by the client in such a packaging scheme, this type of practice has tended to confuse rather than clarify the performance benefits and design objectives of Integrated Building Automation Systems.

At the other end of the spectrum is the fully Integrated Building Automation System, wherein related systems are both physically connective and fully capable of sharing data on both a real-time event basis and a shared data base or historical file transfer basis. Availability of totally integrated systems at this writing ranges from rare to nonexistent. The Building Automation Systems industry has two decades of experience in interfacing dissimilar systems for the purpose of controlling, monitoring, or overriding specific functions from one to the other. These interfaces, however, have typically been performed in external hardware and have been done on a project-specific, custom basis. Although a digital interface between subsystems would be highly desirable and far more flexible over a system's lifetime, few interfaces are accomplished in this manner. In this chapter we will explore some of the driving forces behind this apparent contradiction--market forces and technical performance opportunities that would demand integration, and the yet unanswered questions of how it should be accomplished or by whom.

First, let's examine the market forces or the potential end-user benefits of a fully Integrated Building Automation System, if such a product were to exist. By definition, a fully Integrated System would include physical and functional access to building operations data from any system or subsystem within the building, effectively interfacing all controllers, processors, or data acquisition paths. Current technology would suggest that most, if not all, of these devices are digital in nature and would interface on some sort of data communications link. The list of systems potentially falling under this umbrella is extensive, but would include at a minimum the following:

- Engine Generator or Standby Power Systems
- Uninterruptible Power Supply (UPS) Systems
- Emergency Lighting Systems
- Lighting Control Systems, either discrete or integrated into fixtures
- PBX or Telecommunications Systems
- Office Automation or Management Information Systems
- Control Systems on packaged equipment, including chillers, boilers, computer room HVAC, kitchen equipment, laboratory equipment, etc.
- Fire Management Systems, including detection and smoke control devices
- Security Management and/or Access Control Systems
- Time and Attendance Recording Systems
- Maintenance Management Systems
- Miscellaneous Building Systems, including booster

pumps, sump pumps, etc.

- Elevator Control Systems

Assuming that both physical connectivity and appropriate protocol conversion or standard protocols existed between all of the above systems, a truly holistic approach to building operations could be applied. Experience in Building Automation to date has shown that the more fully connective and integrated these types of subsystems are, the more effectively and efficiently a building can be operated. Some examples of this are included in the Case Studies section of this book.

Take, for example, the potential overlay of an electric load control scheme on such a system. Given the necessity to reduce electrical usage during either a peak demand or time-of-day billing situation, a system with high level access could accomplish the load reduction without significant service disruption or occupant discomfort. A few percent of the lighting load could be reduced by selective reduction of lighting levels at the perimeter through either dimming ballasts or split wattage fixtures, assuming sufficient daylight is available. Elevator speeds could be reduced, or a portion of cars temporarily parked. HVAC system load could be temporarily reduced by incrementing set points rather than wholesale shutdown of equipment. Engine Generator or UPS equipment could be exercised and at the same time contribute to load reduction at the incoming service. Occupancy schedules, either programmed or monitored by the Access Control System, would automatically alter the strategy.

Fire Management functions could also benefit significantly from this type of integration. Readily available real-time access to the control or override of all building systems would allow many functions to take place from a fire fighter's control panel. In addition to receiving alarm and event information from the full spectrum of building systems, fire fighters could readily issue override commands to smoke control systems, elevator systems, electrically locked doors, and auxiliary power and pumping systems.

While these types of integrated functions--either independently or in combinations--have been implemented in current systems, they have typically required a great deal of interfacing and/or overriding of existing controls by the Building Automation or Energy Management System. Typically, then, a subsequent change to the original strategy requires significant changes to either hardware or software.

EXPERT SYSTEMS

Expert Systems can be defined as the capturing of expert logic or operations knowledge in a delivered or installed system. Expert Systems technology has made its way into many businesses and industries, and in fact it is present on a limited basis in existing Building Automation Systems. Optimal HVAC start-up prediction, chiller plant automation, and adaptive tuning algorithms are examples of lower level implementations of Expert Systems currently in use today. As Expert Systems are more widely understood and accepted, software tools will become more readily available for the application of them to real-time problems. Imagine the productivity and performance of a building-wide network whose operations could be directed with the multifaceted knowledge of experts from throughout the user's company or the building operations industry in general. Current real-time computing technology applied to a fully integrated Building Automation System would yield many real and justifiable enhancements to building operation. Add, as a likely future development, the evolution of real-time Expert Systems into the Building Automation designer's tool kit, and the case for full scale integration of Building Automation Systems makes practical and potentially economic sense.

PROTOCOL TRANSLATORS

In the scenario described above, the key element is the ability of systems and subsystems to intelligently interact the way a human would perform the same operations. In other words, anything an operator can do to a packaged HVAC unit, a chiller, an engine generator, or an elevator controller from a control or service panel could be accomplished through a remote digital control link. This requires, of course, some method by which digital processors can talk to each other in mutually acceptable, real-time languages or protocols. At this writing, such a function requires protocol translation between a Building Automation System and the controlled subsystem. This protocol translation can be done in a computer with firmware, simply converting and/or buffering bits and bytes between the systems, or in a software-driven protocol translator most commonly built around microcomputer hardware.

Protocol translation is somewhat more myth than reality. First, it is expensive. One is typically confronted with the challenge of any custom software development project--limited resources, limited time, and loosely defined performance criteria. Since in many cases the ultimate functionality of the integrated system is yet to be discovered, it is somewhat difficult to put boundaries on the capabilities or performance criteria of the Protocol Translator.

Again, opposing forces come into play. The ultimate functionality is usually directly proportional to the amount of effort and expense involved in designing and building the translator. Real-time performance, specifically accuracy and throughput, are very difficult to determine in the early stages of designing and building a Protocol Translator and may, in fact, not be verifiable until the project is complete. If a Protocol Translator is built on a custom, one-of-a-kind basis (as many are), it is almost assured that long-term support and maintenance of the software will be both challenging and costly. It is important to understand the role of Protocol Translators, however, because they are the essence of what most "brochureware" integration is made of. They are cloaked in a variety of depictions and buzzwords, but usually appear as a box on a network diagram called a "gateway," a "network interface," or some other suitably cloudy term. The function and intent, however, is always the same--to make two dissimilar systems interact as if they were the same, with a device in the middle doing the conversions to make either side of the network comfortable.

If protocol translation is an undesirable solution, why do we see it required and/or proposed? Very simply because this solution, although expensive and cumbersome, provides an answer of sorts to the basic need for dissimilar systems to operate in an integrated fashion. The fact that their shortcomings are accepted as a solution gives credence to the assertion that both market wants and performance capabilities of fully Integrated Systems are recognized and accepted.

SHARED COMMUNICATIONS MEDIA

For a while, a considerable mystique developed around the idea of a single network backbone performing all data communications in a building. Again, marketing hype, rather than functionality, tended to be the driving force. While it is true that building-wide communications on a single medium--broadband or baseband cable, fiber optics, or even twisted-pair wire--has a very elegant ring to it, the economics of converting several existing communication interfaces to these types of networks rarely proves favorable. Again, the question "Just what functionality does this integration provide?" generally goes begging. Integration for the sole purpose of sharing communications media is sometimes an economic benefit. It is certainly a more complicated technical solution.

If there is marketplace want and user benefit from fully Integrated Building Automation, then why isn't it more readily available? Well, it's not easy to do. The current state of standards for real-time interchange of data between devices provides a system designer with no logical starting point. Although some basic standards for the physical link between devices such as RS232, RS422, RS485, etc., exist, the protocols or communications data structures between devices have not captured the attention of recognized standards bodies or vendors each pursuing independently their own "best" structures for their own needs.

This is, of course, a historical problem with newly emerging technologies, and the digital communications industry is no different, except that developments and technical progress has occurred at such a rapid pace that standardization is almost doomed to be after the fact. In essence, the designer of a fully Integrated System of any type currently faces a moving target as newer, faster, and more exciting data interchange techniques become available almost monthly.

Second Integrated Building Automation is not cheap. Many users fantasize about the idea of interconnectability between systems of different manufacture. This is generally thought to be in the user's interest, since one could presumably shop for the best price/value combination in adding a feature or function to an existing system. Only rarely is the idea that such flexibility might have a heavy initial cost impact openly discussed. In fact, one reason that standards are so difficult to arrive at in any industry is that each designer or vendor truly and ardently believes that his approach is the most efficient, the most cost effective, or the most desirable for the user. Standardization of interfaces and protocols would probably, therefore, impose some limitations on designers who would find their abilities to optimize cost/function relationships limited by the prescribed communications media and protocol standardization efforts.

On the other hand, the need for standards to reduce life cycle costs should inevitably create a minimal restriction for designers in these areas.

This line of reasoning does not oppose nor rationalize the industry's lack of development and adherence to standard interfaces or protocols, but simply illustrates some of the restraining forces that have kept widespread standards from becoming a reality.

FUTURE STANDARDIZATION STEPS

When and how will higher levels of integration occur within Building Automation Systems? Both "when" and "how" are difficult questions. One thing is for sure. It will not be by some sort of megamerger or "we do it all" strategy by a major vendor. Technologies involved in control systems, telecommunication systems, data communications networks, etc., are each diverse, complex, and rapidly moving in their own industries. It is beyond rational thinking to suppose that one organization is going to be able to provide the most competitive and state-of-the-art offering in all of the electronic system disciplines required for a given facility. Any user should be skeptical of an organization which proclaims to be a "one-source" solution to the entire gamut of technologies involved in building and operating a facility inhabited by a modern business. No one is going to have the best solution across the board at any one time, and most building owners recognize this fact. On the other hand, the emergence of interface standards between systems and devices in the commercial building would provide a responsive answer to the marketplace, while at the same time preserving the necessary technical and marketing specialization within each of the subsystem disciplines. Thus, a marketplace and technical balance is achieved very similar to that existing today in consumer audio and video equipment, where specialty niche vendors co-exist and prosper alongside major system vendors, each complementing the other.

Clearly, this discussion centers around level interfaces for the purpose of digitally sharing data and control signals between major subsystems. It is most likely that these types of standard interfaces would evolve on a "top-down" basis. It is unlikely and probably only marginally desirable that individual sensors, actuators, or terminal devices would become standardized or interchangeable. In other words, a significant amount of performance and functional differentiation would still exist between competitors in the subsystems arenas.

How might standard interfaces or protocols between building systems take place? It is likely that efforts on the order of those being put forth to standardize Manufacturing Automation Protocol (MAP) and Technical Office Protocol (TOP) will be necessary. As a further assumption, then, this charter must be accepted by an existing standards body or must emerge from within the Building Systems industry. It is probably safe to say that a disciplined protocol structure, such as the Open Systems Interconnect (OSI) model as developed by the International Standards Organization (ISO), will form the basis for a successful standard. Nearly all protocols currently in existence for use within Building Systems have been developed by and for vendors considering their use only in proprietary systems. A few vendors have clouded the issue somewhat by publishing their protocols, thereby claiming them to be "open." Most, if not all, of these existing protocols bridge multiple layers of the OSI model and, therefore, will probably preclude their use by the broad spectrum of systems listed earlier in this chapter.

A likely ingredient to such a standardization effort would be the commitment by one or more large users to involvement in and support of a standardization body. General Motors has become a critical ingredient in Factory Automation integration efforts through MAP, and has acted as a catalyst to drive the continuing work on MAP standards.

So much for the future. What about the present? Users desire and are currently specifying functions which require a level of interface between multiple electronic systems within buildings. Uniform interface standards do not exist and probably will not for some time. The essential ingredient in a successful project is, therefore, the ability to integrate, on an engineered system basis, the necessary and essential functions required to make a building work efficiently and effectively. This requirement for effective integration capability is an organizational rather than a product attribute. To be successful in this endeavor, an organization must be chosen that has the necessary engineering and technical skills to design, manufacture, install, and commission combinations of standard and custom hardware or software necessary to meet the design objectives of a project. It is essential also for such an organization to have a great deal of experience in this type of work. In the final analysis, that organization must also have a measure of flexibility and not be committed to a single vendor solution, but rather an "open interconnection" posture. Although high tech systems and skills are being offered as solutions, the basic objective is an efficient, safe, and comfortable building.

Chapter 3

Integrated Energy Management: Individualized Dynamic Control

Thomas Hartman, P.E.
The Hartman Company

Some years ago, the adaptation of microelectronic technology in building controls began the evolution toward Dynamic Control of the heating, ventilating and air conditioning (HVAC) components in modern buildings. Dynamic Control has evolved rapidly in the last few years and now provides two important benefits to building operation that cannot be duplicated by any other method of control: 1) Reduced energy and operating costs for the HVAC plant, and 2) improved and increasingly "individualized" comfort control to the tenants.

THE EVOLUTION OF DYNAMIC CONTROL UP TO TODAY

Deficiencies of Pneumatics Identified

Almost a decade ago, it was recognized that the new microelectronic technologies could be utilized to correct several very basic deficiencies of the standard pneumatic control strategies that had been applied to HVAC control for more than a generation. Traditional pneumatic control strategies were based on "steady state" control theory (steady state control means that space temperature setpoints in a building are held constant by adjusting the HVAC system response to maintain continuous heat flow equilibriums within each space). While the steady state control strategies had served the controls industry well through the era of pneumatic controls, several serious deficiencies were becoming apparent as the introduction of more sophisticated HVAC equipment led to the need for more effective control of system components. The first serious deficiency was the lack of coordinated control among the various HVAC control components. For example, a single building may have several air supply systems, each with a cooling coil, but only one chiller to provide the cooling. No reasonable method existed to coordinate the operation of the chiller with a requirement for cooling by any of the systems. Under pneumatic control, the chiller simply started at a fixed outdoor temperature, or in some cases operated continuously, with corresponding wastes in energy and high maintenance costs.

A second more serious deficiency was also becoming apparent. Although the idea of operating the HVAC system simply to maintain equilibrium heat flow conditions within each building space was a mathematically correct means of maintaining space temperature setpoints, the actual application of that principle in the newer HVAC systems often led to comfort and energy consumption problems. In warm weather, systems with full economizer modes still could not utilize extra cool outside air during the morning hours to flush out the building and reduce the need for mechanical cooling later in the day, and in cold weather, the low fixed supply air temperatures from VAV terminal units made building warmup difficult and inefficient, and often created uncomfortable drafts in the spaces during the day.

Microelectronic Technology Key to Solutions

Those of us working in the industry at that time saw that new microelectronic control systems could be developed to solve these energy and comfort problems. As a result, a new approach to HVAC control has evolved that uses the speed and calculating capacity of the microelectronic-based control system to make intelligent decisions and step beyond the limitations of the old pneumatic steady state control strategies. This new control strategy has become known as Dynamic Control and it is now in use in many buildings across North America today.

Dynamic Control Characteristics

Dynamic Control is the name for this new building environmental control approach because it operates HVAC equipment by anticipating time-based changes in heat flow patterns within the building, and in so doing eliminates the need for the continuous heat flow equilibrium conditions of pneumatic steady state control. Dynamic Control focuses directly on comfort and uses the intelligence of its anticipatory routines to manage disequilibrium heat flow conditions in order to provide comfortable conditions with greater energy economy and less possibility of local temperature variations. Dynamic control strategies recognize that because of the ever changing heat flow patterns within any commercial building, continuous equilibrium conditions are in fact neither desirable nor even possible in any practical sense.

In addition to the substantial comfort improvements, there are enormous energy savings opportunities available to buildings employing Dynamic Control. Utilizing coordinated control and managed disequilibrium concepts, Dynamic Control has been successfully applied to buildings through the energy crisis years when its major emphasis was energy reduction, all the way to the present when that focus is now shared with comfort enhancement, local temperature control, and simplified building operation.

The strength and effectiveness of Dynamic Control strategies depend largely on how effective the software elements (or subroutines) work independently, and together. Emphasis needs to be placed not only on anticipatory routines such as the outdoor temperature projector and the building thermal inertia evaluator, but also on the algorithms that interpret this and other data to establish the real time setpoints and control decisions for output to the HVAC equipment. As a result of particular building configurations and individual engineering emphasis, there are always several possible variations in Dynamic Control strategies for any building, but to be considered a truly "Dynamic Control" approach, two essential ingredients must be present:

1. Fully integrated control: All HVAC components must be made to operate in a coordinated manner. Each must operate only when actually required and in coordination with all other components to insure energy waste does not occur from overlapped, insulated, or non essential operation.

2. Intelligent Anticipatory Control: All HVAC components must be operated with consideration that building heat flow factors change with time. The system must establish setpoints and operate equipment in anticipation of these upcoming changing conditions.

These two features of Dynamic Control constitute a very condensed statement of what is particularly new and exciting in the application of computer-based Dynamic Control strategies. While this particular technology continues to be further developed and improved, it is now well proven in its many installations to date.

BAS Requirements for Dynamic Control

Applying Dynamic Control strategies in any building requires some very specific Building Automation System (BAS) features that are not available with all computer-based systems. The most crucial features required are:

1. A powerful BAS system that is designed with sufficient input points, computing capacity, and processor accessible files to collect and process data for the required global decisions regarding coordination of operation and anticipation of upcoming conditions.

2. A flexible BAS general control language that allows special calculation and control algorithms to operate effectively.

3. Direct digital control (DDC) of each HVAC output point tied directly to the general control language decisions.

Benefits of Dynamic Control

Once Dynamic Control is in place, the benefits are immediate and substantial. With Dynamic Control any building will experience the following benefits to a degree that depends on the building envelope, HVAC system characteristics, and building location:

1. Improved Energy Performance: The most important benefit of Dynamic Control; it provides a strong economic incentive for its use. Very substantial energy cost savings can be expected in all applications.

2. Improved Comfort Conditions: Resulting from the anticipatory aspects of Dynamic Control. Comfort improvement is nearly universal to all Dynamic Control applications.

3. Reduced Maintenance costs: A spinoff of reduced equipment operating hours and are significant in many projects.

Though we all see some modern computer-based energy management systems installed without improved control strategies, Dynamic Control continues to build in popularity. There are now a number of different names given Dynamic Control such as "feed forward," "adaptive," or "anticipatory" control. The term Dynamic Control is generally preferred because it best contrasts the difference in performance capabilities between pneumatic and computer-based control systems.

NEW AND EMERGING DYNAMIC CONTROL CONCEPTS

As discussed in the first part of this book, the concept of an intelligent building actually denotes a continuum of capabilities. The forward end of that continuum is always being redefined by the state of microelectronic technology and our understanding about the human needs for which buildings are constructed. At present both are moving forward at breakneck speed. In the field of HVAC controls, the industry is now in the midst of a giant leap forward that is opening new possibilities of further advances for HVAC control capabilities in the intelligent building.

The Terminal Control Revolution

In the past, virtually all computer-based energy management systems have provided control for only the central HVAC equipment while relying on traditional pneumatic or electronic controls for the local control of terminal equipment. Now building automation systems are beginning to offer competitively priced control packages that permit the integration of terminal equipment into the BAS. With this control capacity now a reality, the HVAC control revolution that began with the introduction of Dynamic Control nearly a decade ago is spreading out to the control of terminal devices and changing that segment of traditional heating, ventilation, and air conditioning operation even more profoundly than it has changed the control of central systems.

The reason the present changes in terminal control are so significant is twofold. First, the old means of terminal control, usually by area thermostat control, was never very effective in providing individual worker comfort. Too often the thermostat was poorly located or influenced by unusual local heating or cooling loads that were not typical for the larger space it was to control. The thermostat was also never able by itself to compensate for radiation changes due to sun or cold perimeter walls and glazing. Second, the continuous addition of heat generating components to individual workstations is causing ever increasing problems for these pneumatic-based terminal control schemes. Simply stated, the wall thermostat method of terminal control has already been out of date for some time, and now is losing its purchase cost advantage. As the connection of terminal HVAC equipment to the building automation system is becoming available, a radical change to the concept of intelligent terminal control is occurring with three schools of thought:

1. AREA TERMINAL CONTROL (ATC): The ATC approach envisions replacing all pneumatic thermostats with electronic temperature sensors that are tied to the building automation system and programmed to provide fully automatic local environmental control.

The ATC concept relies almost entirely on the intelligence of the building control system to interact with the space conditions and provide suitable comfort conditions with no direct means for occupant reset. The ATC concept is the most direct route to incorporating terminal control into the building automation system, and is the scheme that is presently favored by most in the industry.

2. LOCAL TERMINAL CONTROL (LTC): The LTC approach is similar to the ATC approach except that it provides some means of local environmental control from the occupants (most common is a warmer/colder adjustment knob on the space temperature sensor that replaces the pneumatic thermostat). This approach operates much the same as the pneumatic thermostat of the past except the LTC approach allows additional overall supervision from the building automation system. The LTC concept is closest to using the building automation system to simply replace the pneumatic terminal control networks with similarly operating control loops (though the LTC concept can also be centrally reset).

3. INDIVIDUALIZED TERMINAL CONTROL (ITC): The ITC approach is an entirely new concept that has been made possible by the new microelectronic control technologies and our understanding of the factors that influence human comfort. The ITC approach allows individuals at their workstations to provide adjustments to the comfort conditions at their location. This concept utilizes HVAC components installed directly inside or adjacent to each individual's workstation. Some means of individual temperature and air flow control adjustment is provided directly to each building occupant under this concept.

The Future of Terminal Control In Intelligent Buildings

Each of these three terminal control schemes has its supporters, but recent tests in the US and Canada seem to indicate that while one would expect the ATC approach to be the least expensive and the ITC approach to be the most expensive, that is not necessarily the way it will will work out. In many applications the ITC approach may actually be less expensive from both first cost and operating cost perspectives. Any developer or design professional involved in a building project must therefore look very closely at these three approaches to terminal control and the goals of the building function before jumping to any conclusions about the cost effective means of terminal control. The salient issues regarding the best means of terminal control and the reasons the ITC concept looks to be the terminal control mode of the future are:

1. Individual Environment And Productivity: Employees are demanding better environmental conditions during their workday. Meanwhile, the proliferation of electronic office equipment, because of its concentrations of heat and equipment ventilation fans, is making this an increasingly difficult job for the building HVAC system to accomplish. The ITC concept envisions that the HVAC terminal equipment and an operator interface will be integrated in, or adjacent to, the workstation where the operator can make adjustments for maximum individual comfort.

2. Distribution of Ventilation Air: Because the terminal outlet is workstation-based in the ITC concept, the operator will never be in a stagnant area, even if the office layout requires a high concentration of workstations in some areas. Also, the air cleaner within the workstation would reduce the carryover of smoke or other odors.

3. First Cost and Operating Cost Considerations: In many areas of North America, the first cost of an office building is actually lower with the ITC concept than either of the others because ITC utilizes a far simpler central HVAC system whose

reduced cost in many instances will more than offset the increased cost of the ITC terminal units. Furthermore in a increasing number of building occupancies (those where the employees operate under a flextime schedule, for example), the operating (energy) costs of the ITC concept will be significantly lower than either of the other two concepts because the workstation-based terminal device operates only when the occupant is present. The operating costs associated with the ITC concept will also be lower than other terminal control concepts because a more effective mode of Dynamic Control for the central HVAC equipment can be employed.

4. Reconfiguration Flexibility: Because the ITC concept envisions terminal equipment as a part of each workstation, periodic revisions in office layout (the first revisions often begin even before the building is occupied) will not require the special attention and rework of the HVAC distribution equipment as these systems do today.

5. Investment Considerations: Under the ITC concept, the terminal units will normally be owned by tenants as part of their office equipment. This has many advantages over leasehold assets, including depreciation advantages, US investment tax credit, and a capital life beyond a single lease agreement.

The Future: Individualized Terminal Control (ITC)

In view of these considerations, it is clear that there are many buildings for which the Individualized Terminal Control (ITC) concept would have a clear advantage over the other integrated terminal control concepts. However, for the next decade, there will be one offsetting disadvantage for the ITC concept: it is new! Because it is new, there is not a great deal of experience with the hardware components of the ITC scheme, or the human requirements of the system. One strong driving force for the development of effective ITC-based intelligent buildings is that the ITC concept allows "individualized" environmental comfort levels. In so doing, ITC recognizes that there is not a single suitable comfort level that satisfies all of today's building occupants all of the time. It usually is this reason more than any design deficiency that is responsible for comfort problems in buildings today.

Based on development and testing work that has already been done, the form of the ITC units in buildings of the near future will include the following characteristics:

1. The terminal unit will be contained in the workstation.

2. A hybrid heat pumping and radiating scheme will interact with the primary air to maintain individual comfort.

3. Simple controls will react to both operator and computer system instructions.

4. The ITC unit can be either directly or indirectly connected to the central air system.

5. The ITC unit will permit special air cleaning and smoke removal accessories to meet specific individual comfort needs.

While information accumulated to date does point to a very definite direction for the future of ITC-based systems, a great deal of research and testing will be required to fully develop the ITC concept. It is important for firms and organizations who will be affected by ITC systems to get involved now with their development. In particular, the following groups should take an interest right now:

1. Terminal Unit and Office Furniture Manufacturers

2. Architect/Engineering Firms

3. Building Owners and Developers

4. Associations, Foundations, Government

5. Building Automation and Control Companies.

The future of building automation systems in intelligent buildings is an exciting and rapidly evolving one. The present trend to extend Dynamic Control strategies to the terminal delivery units of HVAC systems is an advance in technology that truly has something for everyone. For developers it will mean simpler and lower first cost buildings. For tenants it means lower energy costs, improved worker productivity, and far greater flexibility in space utilization. Finally, for office workers it means the opportunity for individualized comfort conditions and improved ventilation control. A lot of questions need to be answered about integrating terminal control into the building automation system, and the three alternate integration concepts, but with desirable benefits like these, it is certainly not too early for building design professionals to be considering these terminal control concepts for incorporation into buildings that are now in the design stage.

Chapter 4

Fire Management:
Intelligent, Integrated, Informed

Ralph E. Transue, P.E.
Rolf Jensen & Associates, Inc.

Everyone agrees that fire can take a heavy toll on people, property, and continuity of operation...that it can, in addition, have side effects ranging from negative publicity to crippling liability exposure. An effective fire safety program is therefore critical for buildings of any size, and any level of sophistication.

The basics of fire safety apply to every building - intelligent or not. (See Figure 4-1)

But the advent of the intelligent building has opened up a range of possibilities for those of us concerned with creating safe environments for building occupants. It has, in fact, given us new opportunity to optimize the elements of fire safety.

That is partly a result of the march of technology, and partly the result of sticking to basics. Indeed, in the pages that follow, we will explore a number of the physical components of fire control and information management.

Keep in mind, however, that their value will be directly proportional to the effectiveness of the less tangible elements of fire control:

- Planning, starting with a fire safety program as an integral component of the building concept -- a component to be incorporated in the architects' and engineers' detail design, and carried through in every phase of construction and installation.

Fig. 4-1 Firesafety Concepts Tree

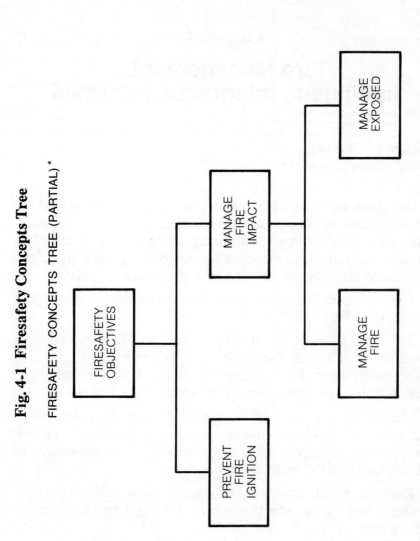

FIRESAFETY CONCEPTS TREE (PARTIAL) *

* For additional information and a more complete concept tree, see NPFA 550-1986 Edition, Guide to the Firesafety Concept Tree, National Fire Protection Association, Quincy, MA

- Tailoring every aspect of the program to the particular requirements of a given building.

- Commission testing, including activation of all components of all systems including interfaces and control functions on an individual basis, before the building is occupied -- and on-going testing thereafter.

- Intelligent management, directed by an operator who understands both the individual components of the system and the manner in which they work together to meet the fire safety program's objectives.

- Continuing education of both employees and occupants.

- These elements are every bit as important to fire safety as the actual physical components of the systems involved. A successful fire safety program incorporates these elements from the start.

The goal of the fire safety program is to provide arrangement of fire control elements which operate effectively together without excess.

Among the questions the program must address are these:
- What code requirements apply?

- Can building design objectives be achieved within code requirements? What alternatives exist to meet the intent of the codes?

- Will an automatic suppression system be installed?

- What form of containment will be designed into the building?

- How will people be relocated, evacuated or defended in place?

- What forms of detection, alarm and communication

systems will be needed?

- How will fire fighters approach and enter the building in an emergency? What are their requirements?

- How will fire products be managed?
- Who will be the users of the systems? What are their needs?

There are, of course, no stock answers to these questions. They must be addressed within the framework of the individual building's unique design.

The Components of Control

Fire control requires some combination of the following elements:

- Containment via spacial separation or physical barriers
- Extinguishment by automatic and manual means
- Alerting of fire fighters, security staff and occupants
- Disposing of such fire products as smoke, gases and heat

Generally speaking, these tasks will be handled by systems falling into one of two categories: passive or active.

Passive Control Systems

Passive fire management components such as fire walls and smoke barriers may not receive much attention in intelligent buildings; they are not as glamorous as systems for building automation, data transmission and shared tenant services.

But when fire strikes, they are hardly dull: They spring to life as active containment elements that enhance safety and make the challenge of extinguishing fire far more manageable.

The trouble with fire walls is that, no matter how well conceived they may be, they do not always remain as effective as a building's designers intended them to be. The designers may have provided for a veritable fortress of two-hour rated fire walls* to contain hostile fires. But the sad fact is, occupants too often begin violating the integrity of those walls as soon as they move in by punching holes through them for cables, wires, ducts and pipes.

The solution: Designing and installing fire walls properly, and then operating the building in a manner that maintains their integrity. That means eliminating the <u>need</u> for adding new cables, wires, ducts and pipes.

That is one area in which the intelligent building offers the potential for superior fire safety: an intelligent building can be designed with adequate data highways, communication channels, control circuits, and mechanical utilities to meet occupant needs.

Of course, the utility highways themselves must include fire resistant properties, and must be properly sealed wherever they penetrate fire walls. The points are simply these: Building code requirements for fire-related partitions do not have to restrict the use of utility highways in intelligent buildings. More importantly, by providing adequate utility highways for occupant communication and comfort needs, the designers of intelligent buildings increase the likelihood that the building's containment systems will remain at a high level of integrity.

Passive control elements deserve careful consideration in the earliest stages of building design. For building designers, passive elements include:

- fire walls
- smoke barriers
- spatial separation
- exposure control

*The integrity of building partitions is measured in terms of an endurance of rating in a standard time-temperature test. See Fire Protection For the Design Professional edited by Rolf Jensen, Boston, MA: Cahners Books, 1975, 197 p, includes index.

- exiting capacity
- fire service access
- selection of materials

For building operators, they include:

- emergency planning
- housekeeping standards
- staff training programs
- fire prevention
- selection of furnishings

Active Control Systems

Active control systems can be divided into these basic categories:

Automatic fire suppression, including:

- automatic sprinkler systems
- special hazard systems

Manual apparatus, including:

- standpipes
- hoses
- portable fire extinguishers

Information management, including:

- emergency communication to occupants
- 2-way fire fighter's communication system
- fire command station
- fire detection/alarm system

- supervision of suppression, detection, alarm and communication systems

Control interfaces to other systems, including:
- air handling equipment
- elevators
- door control

Clearly, there are many fire management tools available to today's architects, engineers, developers and building operators.

The challenge is deciding which ones to incorporate in the fire safety program for each unique building.

As an example, consider the question of how extensive the smoke management systems should be as a part of the fire safety program for a building. When the fire is small, its products -- heat, smoke and gases -- are limited to its immediate vicinity.

As a fire grows,

- heat impinges on a broader area
- energy is released at a increasing rate
- smoke is produced in ever increasing volume
- temperatures rise
- the severity of the threat increases

It is apparent that the earlier the fire can be extinguished -- preferably by automatic means -- the more limited the damage and threat from these products will be.

Smoke management can be accomplished by an arrangement of air handling systems and smoke barriers to dispose of fire products and limit their spread. Together, for small fires these systems can:

- Remove smoke from the fire area via negative air pressure, and

- Prevent its spread via positive air pressure in areas adjacent to the fire.

Smoke management systems can also be effective at distances from the fire -- away from the high energy region -- for somewhat larger fires. However, a hostile fire burning freely will grow to the point at which the volume of smoke and gases produced (cfm) and the rate of energy release will be such that mechanical systems will be overwhelmed.

The design goals for including smoke management systems are to provide clear air in evacuation routes for occupants, and make it easier for fire fighters to approach the blaze.

It can be argued that a building equipped with a complete automatic suppression system covering all areas does not need separate smoke management systems. One thing is clear; a properly designed, installed and maintained automatic suppression system will limit the smoke produced by the fire; either by extinguishing the fire or halting its growth. No similar claim can be made for smoke management systems. While smoke management systems may contend with the products of a small fire, they do nothing to limit the growth of the fire; nor control the source of those products and their threat.

The fire safety program must consider questions and comparisons such as these, as well as code requirements, to result in a functional and effective program without unnecessary systems or unnecessary complexity.

It is critical to address these issues up front, while developing the building concept to assure that the fire safety program is reflected in the architects' and engineers' detail design...and to see that its components are installed properly throughout the course of construction, by everyone from carpenters and masons to electricians and pipe fitters.

Information Management/People Management Systems

The electronic systems and communication capabilities of intelligent buildings provide designers with a tempting array of possibilities for management of information and people in a fire emergency. The electronics revolution has given us a steady flow of new capabilities including:

- computer-based alarm systems
- multiplexed wiring
- digital communication

Developments such as these have created an opportunity for interfacing or integrating a building's fire safety systems with other building control and communication systems.

But these developments have also created new responsibilities for the designers and installers of these systems...to see that the systems operate in a coordinated and compatible manner.

When the concept of integrating these systems was introduced in the early 1970's, building automation systems were quite expensive. The proposals to integrate fire alarm and supervisory system functions into other building control systems were seen as a way to enhance cost-efficiency of the building automation systems.

Since then, advances in microprocessor technology have reduced the costs of control systems dramatically. Today, there is little financial advantage to integrating these systems. As a result, designers of the intelligent building would be wise to maintain dedicated fire safety systems, connecting them to other building control systems only to the extent required for fire protection purposes in conformance with the fire safety program.

Similarly, voice communication is probably best handled as a dedicated, electronically supervised system to communicate people management information to building occupants and staff.

All systems should meet National Fire Codes, local codes, and be listed by Underwriters Laboratories or an organization that maintains periodic inspection of production of listed equipment and whose listing states that the equipment meets nationally recognized standards for the intended use.

Compatibility of Systems

Passive and active fire control systems do work together. For instance, the presence of fire walls helps fire fighters "knock down" a fire with water streams from hose lines.

In fact, fire walls increase the effectiveness of automatic fire suppression systems as well.

Active systems should be designed for compatibility and mutual effectiveness without undue complexity. For instance, fire detection system zones should be designed to be compatible with smoke barriers if the fire detection system is to initiate a smoke management system.

The fire safety program should be developed with an additional objective in mind: to assure compatibility among systems. The various systems are shown on different sets of design drawings: architectural, mechanical, electrical; they are installed by different trades: mason, pipe fitters, electricians; only by adhering to the fire safety program and inspecting the construction against the plan can it be assured that the designers and contractors have achieved the desired result: compatible systems.

The Essentials

Both active and passive systems require careful pre-planning. They also require operation and supervision by intelligent operators once the buildings are occupied. Both the staff and occupants should be trained in fire safety procedures.

Just as important are testing and maintenance. Before occupancy, a building's fire safety and control systems should be thoroughly commission tested, including activation of all devices, systems and control functions. Electrical verification is important; but it is no substitute for actual operation of all functions.

Summary

In summary, every intelligent building should have a customized fire safety program as an integral part of its conceptual and detail designs -- a program to assure that all the elements will ultimately work together for efficient, effective fire and people management. This plan must allows for:

- Individual tailoring to meet the building's unique requirements. Cost-effective approaches to meet building code requirements and to provide compatible systems:

- Complete commission testing to assure operation under emergency conditions

- On-going testing and retesting of all fire safety systems

- Smart management with a thorough understanding of the fire safety program's role in occupant protection

- On-going employee and occupant training

- Vigilance relative to changes in building use and hazards

By adhering to a properly developed fire safety program and designing into the intelligent building adequate utility highways and providing for fire suppression, the designers, contractors and operators of the building can provide a high level of fire safety.

Chapter 5

Integrated Security Management

Thomas F. Raymond
The Stroh Brewery Company

When measuring building intelligence, it is important to rate the effectiveness of the security management program that protects people and property in the building. Buildings with a high-level utilization of sophisticated electronic equipment that complements security staff members are those with the best program and, therefore, are the most "intelligent."

At The Stroh Companies' world headquarters at River Place in downtown Detroit, a systems approach is used so that electronic security equipment provides an additional level of intelligence to the overall security management program. The systems approach has been effective and will continue to thrive as new and unforeseen needs develop in the future.

The systems approach to security has teamed highly trained security professionals employed by Stroh, by a community security service, and by the Detroit Police Department with sophisticated electronic security provided by a computerized facilities management system.

In developing a systems approach to security and loss prevention, one objective has been to bring both the private sector -- the business people and the residents of River Place and surrounding areas -- together with the public sector -- police and other government agencies -- to reduce criminal activities.

Fig. 5-1   An Artist's Rendition of Stroh River Place Project, Detroit, MI

Stroh River Place Needs Intelligent Building Security

Stroh River Place is a 31-acre, multi-use, planned community located within Rivertown, Detroit's new warehouse district redevelopment project. The Stroh Brewery Company, brewers of nationally-recognized Stroh's beer, has its corporate headquarters at River Place.

River Place is located on the Detroit River, a busy international waterway, less than a mile from the downtown business district with its shopping, entertainment, sports, recreation, and famed Renaissance Center. The late 19th and early 20th century industrial buildings under rennovation as River Place once housed Parke-Davis and Company. The unique buildings are being restored to enhance the architectural beauty of the past and modernized to meet the demands of businesses and residents today and in the future. (See Figure 5-1)

In addition to The Stroh Brewery Company and Stroh River Place Properties, major tenants of River Place include Michigan National Bank and Doctors Hospital. Stroh office personnel might be present any hour of the day, any day of the week. The need for after hours security will increase as more tenants move into the complex.

Office, retail, restaurant and residential space are being completed along with a 1,250-car parking structure. A number of surface parking lots also are part of the complex. Long-term plans include hotel and fine arts structures as well as additional renovation for office, retail and residential units.

The riverfront and warehouse district ambiance, along with the convenient location near major streets, expressways and downtown, creates a cosmopolitan atmosphere that attracts up-scale tenants -- if they can be assured of the security and safety of the area. Studies showed that, along with parking, security was the most frequent concern of tenants contemplating a move to the urban center. Stroh River Place addresses those concerns; its level of building intelligence provided by its security management program is key to its success.

Security System Teams People with Electronics

Prospective tenants' concerns about security were given top priority in planning the River Place project. A security system concept was developed before equipment was chosen. The basic concept was to reduce crime and the fear of crime through an integrated security system in which electronics complemented people, a system in which security officers and electronic equipment worked as a team held together by a set of concepts and procedures. People using an electronic system must understand it and be comfortable with it.

From the outset, the Stroh insurance brokerage company was involved in qualifying needs and developing security system specifications. A consultant from the firm is an important member of the River Place planning team. He participated in visits to vendors who met electronic system specification criteria and to working installations of their systems throughout the United States. He assisted in ascertaining vendor compliance with national standards such as those established by Factory Mutual Insurance Company, Underwriters Laboratory and the National Fire Protection Association. The added perspectives from the insurance consultant occasionally changed specifications as concepts were refined to become reality.

Insurance carriers who bid on his client's coverage look at property protection to determine their degree of risk. The security system specifications play a key role in determining acceptance by insurance carriers and better rates from them.

Integrated Electronic System Specified

Security system specifications for Stroh River Place required an integrated electronic system with flexibility, capacity for expansion, and ease and availability of service and maintenance. Additional criteria included the capability to integrate the nine Stroh plants located throughout the United States to the sys-

Fig. 5-2 The security control center at Stroh River Place is the heart of an extensive security management system that teams sophisticated electronic equipment with highly trained security officers.

tem at the Detroit headquarters and the availability of service at those locations.

The overriding factor in the final selection of the electronic system vendor was that it had trained, qualified people in Detroit who were able and willing to work with those responsible for security at Stroh to keep the system concept intact. Also, they were able to offer voluntary alternates for implementing the integrated system within the budget established by Stroh.

The electronic security system selected is a facilities management system with a fully integrated network of computers installed throughout the 31-acre Stroh complex. Each is interconnected by fiber optic cables to provide the utmost degree of reliability and long-term flexibility. In its entirety, the computer network provides sophisticated fire and security management, card access protection, closed circuit television monitoring and communications as well as energy management capabilities. The facilities management system is Underwriters Laboratory listed and Factory Mutual approved.

Security Control Center Developed For System

Some facilities management systems, no matter how sophisticated they are or how large an investment they represent, are accommodated in less than ideal environments. Space is found for the central processing unit and primary monitors, keyboards and printers after the system is installed, occasionally in inappropriate locations adjacent to mechanical equipment areas where cramped conditions as well as dirt, noise, vibrations, heat and humidity negatively affect the equipment and its operators.

Planning for River Place included a special security control center to accommodate the facilities management system and the needs of the security officers who would operate it at the security headquarters building. (See Figure 5-2)

Nationally-recognized suppliers along with the security center operational staff were involved in the center design. Because of this integrated design work, the main consoles of the system are housed in a large computer room with raised floor, a halon system, and close control of temperature and humidity. Video display terminals and printers that provide information about card access, security, fire and energy management as well as telephones, intercoms, closed circuit television monitors and other essential equipment are accessible to officers using the consoles.

An additional level of building intelligence provided by the electronic system allows the center to monitor and control operations at plants outside Detroit by using modems to communicate with remote stand-alone computers over telephone lines. Specifications required that telephone line communication would be provided with necessary compatibility.

In addition to the computer room, the security headquarters building contains a strategic planning center, an equipment room where some maintenance is done, an operations room where photos are taken and badges for the card access system are processed and issued for employees and residents, and a training room. The center also houses the headquarters for the private security service that Stroh helped to establish, Rivertown Security Service, and the east side mini-station for the Detroit Police Department, which Stroh rents to them for $1.00 per year.

Stroh helped to create Rivertown Security Service, with the cooperation and support of the businesses and residents of the Rivertown area, to add another ring of security around River Place. The service is a non-profit security agency with both mobile and fixed posts in the Rivertown area.

The objective of the security service is to bring both the private and public sectors together to reduce crime and the fear of crime in the area. The service is supported by subscriptions of businesses and residents and operates only within the five-square-mile defined area called Rivertown.

Rivertown Security Service began operation in April, 1984. It helped to reduce crime in the Rivertown area by 50% that year and Detroit Police report a continued curtailment in area crime.

Security officers at Stroh River Place are college-trained specialists. Their average age is 31 and all have at least three years experience at Stroh and have been part of the systems approach since its inception. Security personnel monitor the site around the clock and are only a telephone call, an intercom, or a camera away from people as they move about River Place. Stroh employs 22 security officers who work closely with the 15 officers of the Rivertown Security Service and officers of the police department mini-station. These numbers are growing as River Place and Rivertown develop.

Customized Training and Support Important For Success of Program

Although modern electronic systems increase operator effectiveness through "user friendliness," it also is essential to provide ongoing training about the system as well as about other aspects of the security management program. The Stroh Companies have made a strong commitment to training security officers by contracting with the vendor of the facilities management system to provide six hours of training each month for each officer.

The vendor developed a structured training program that has responsibility and accountability for participants and maintains a training program data base that includes proficiency goals as well as other records for each officer. The training room at the security control center is used by the system vendor for multi-media group presentations. Hands-on experience with the computerized system is an important part of each officer's training.

Another example of vendor support is the vendor's ability to perform routine and specialized maintenance of the electronic equipment to insure that the level of building intelligence is maintained. In addition to a typical service contract for maintenance of the computerized facilities management system,

Stroh also required the system vendor to provide specialized maintenance for electronic equipment subjected to a construction environment. For example, smoke detectors require periodic cleaning to function properly when exposed to the considerable amount of dust from renovating the old buildings of River Place.

Fiber Optics Provides Expandability and Reliability

Originally, the electronic system was specified to use copper wire requiring two four-inch conduits between buildings. In bid response, the system vendor suggested fiber optic cabling as an alternative. Fiber optic communication could be installed at no additional cost since extensive conduit could be eliminated. Plans for expansion of the system would have required additional wiring in the future, but fiber optics can handle most planned additions to the system.

Other benefits of fiber optic communication include minimizing interference from lightning or electromagnetic interference, ease of maintenance, and long-term growth potential due to broadband multiplexing capabilities. Furthermore, the distances between closed circuit television cameras and monitors approached limits for good resolution when using copper wire. With fiber optics, there is no distortion on video screens.

The use of dedicated fiber optic cabling and distributed processing technology make unlikely a loss of communication between the control center and the electronic system components in buildings throughout River Place. Dedicated network controllers in each major building of the complex provide stand-alone fire, security, card access, and energy management applications if communication is lost with the security control center.

CCTV Increases Efficiency

Many technological advancements are included in the electronic security components of the facilities management system at Stroh River Place. State-of-the-art technology is especially evident in the extensive closed circuit television (CCTV) network. (See Figure 5-3)

Because most calls for security assistance are informational requests, labor costs can be reduced substantially and availability of officers in emergencies can be assured by careful incorporation of CCTV into the security management system. CCTV cameras have been designed to function with many intelligent, human characteristics at River Place.

Each camera's ability to "see" was extended by specifying user-programmable motion detection capability. Monitors for

Fig. 5-3 System Configuration

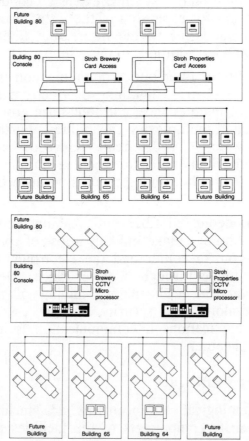

individual cameras are programmed to sound an alarm at the security control center should anyone enter the "seen" area. During the alarm condition, officers at the center check the monitor to confirm the situation and take appropriate action.

Officers can communicate with the person or persons in the affected area because CCTV cameras also were given the ability to "hear" and to "speak" by installing two-way intercoms at each camera location. People at the camera location can summon control center officers by pressing a button on the intercom which also sounds an alarm at the center.

In addition, the camera "remembers" because the CCTV system is programmed to automatically record events on the screen when an alarm is sounded. These video recordings can be valuable for proof of details about an incident.

For this application, CCTV cameras were specified to include a 2/3 inch tube and operate from a power source of 100 to 130 volts AC at 60hz. The cameras provide at least 700 lines horizontal resolution with a dynamic light range of 500,000:1. Mounting hardware plus environmental housings appropriate to individual applications were specified.

The present CCTV system at River Place consists of 25 interior cameras, 27 exterior cameras and 24 high-performance monitors of which four are remotely located. The areas of viewing include hallways, stairwells, the parking deck, elevators and other sensitive areas throughout the complex. The system will expand as the complex expands. Since expansion requirements could not be completely defined in specifications, modular hardware was chosen along with the fiber optic multiplexing for related expandability.

Card Access Control Adds More Intelligence to System

As the security system at Stroh River Place was developed, it was evident that controlled as well as convenient accessibility to areas within the multi-use complex was needed. Integrated card access control sub-systems were specified to meet this need.

Cards are programmed to allow holders access only to needed areas and only during allowed times. As residential units are occupied, many tenants will use their access cards to enter their homes as well as their offices and the parking garage.

When a card is used for an unauthorized area, an alarm sounds at the security control center and officers can take action if needed. Information about card uses is available on the video displays of the system as well as in reports from the system printer. Officers are able to get a printout, for example, of all whose cards were used in any given time period if there is a need to investigate an occurrence in an area.

Officers who need to locate a card holder can call up information on the system screen to learn where that person last used his or her access card in order to quickly contact the employee or tenant.

Ninety card readers are presently installed at Stroh River Place and more are going on line. For added security, the electronic card access sub-system employs Wiegand technology and has a capacity which can be expanded to 256 readers. It can handle 14 remote sites with precision access and time and attendance software. Each card reader has an associated door alarm.

Access control at Stroh's Memphis plant is now being installed with communication to the Detroit central system. Remote card readers are programmed by security officers at the headquarters where personnel records are kept. Specifications for plants in Winston-Salem and elsewhere will be enacted and it is expected that systems will expand as plant management recognizes the efficiences possible with card access control. An example is that time cards could be replaced with capabilities of the card access system. Another use may be starting and stopping process control equipment with card readers, thereby limiting operation to authorized personnel as well as having a record of equipment user.

System Integrates Fire Protection Features

Fire protection features are an important part of the total security system at Stroh River Place. During a fire emergency, Stroh must be sure that personal safety can be maintained with minimal impact to occupants of areas which may not be in immediate danger. System integration makes this possible.

Fire command stations with audio control consoles and fire detection panels are installed in main buildings in the complex. They provide signalling power for one-way voice and alarm speakers in addition to supervision and control of fire alarm speaker zones. If a fire occurs, the system will sound an alarm throughout the building. Firefighters can use the command stations to pinpoint alarms and make announcements to building occupants.

The dedicated network controllers in buildings will continue to function with no loss of programming features or functions if communications are lost with the security control center. So too is each fire field processing unit equipped with circuitry that allows it to operate on a stand-alone basis in the event that communications to its dedicated network controller are disrupted.

Fire field processing units have locked doors with see-through viewing areas for alarm and trouble annunciator lights. In the event of a power failure, battery back-up provides an additional 24 hours of function for each unit. Multiplex trunk transmission to the units is redundant according to National Fire Protection Association 72D, Style 7, E. The redundant trunks are designed to allow the system to automatically switch to standby transmission in the event of a trunk failure. The secondary trunks have monitoring integrity equal to the primary ones.

Fire alarm speakers are UL-listed for emergency voice warning and communications. Speaker cones are fire retardant and moisture repellant. Also, transformers can be adjusted to suit the ambient noise level in the areas where they are installed.

Fire alarms can be activated by a number of devices. Tamperproof, red, cast metal pull stations are located throughout the complex as are smoke detectors and sprinkler water flow switches. All will sound an alarm at the security control center as well as in individual buildings.

Ceiling-mounted smoke detectors are solid-state, photo-electric types, listed by UL and approved by FM. Duct smoke detectors are similar and also incorporate sampling tubes across the complete face of the ductwork of which they are part. Remote alarm lamps allow the rapid location of concealed smoke detectors in alarm.

In the event of a fire alarm, door strikes unlock, releasing all preprogrammed doors. Fans are activated to exhaust smoke from the building affected. Lights in the fire area that have been turned off after hours for energy savings are automatically turned on to further aid occupant egress.

To evaluate the fire and security program and look for potential problems such as devices that are temporarily off line due to construction in progress, managers are able to print logs and summaries of equipment connected to the facilities management system. Later, this reporting capability will be used to streamline maintenance operations by indicating need for periodic testing or service.

Elevators Have Fire and Security Features

Building intelligence is enhanced by fire and security features added to elevators. Smoke detectors are installed in each elevator lobby. In the event of an alarm, elevators are recalled to the ground floor or to an alternate floor if there is smoke on the ground floor. Elevators then are available for use by firefighters. A manual override switch that simulates the smoke detector recall is installed in lobbies so that officers can recall elevators to the ground floor if needed.

All elevator cabs are equipped with special intercom stations which also can be monitored by the security control center. If an emergency call button in an elevator is depressed, officers at the center see a color graphic display detailing the exact location of the elevator car and the response that is expected to be taken. Also, reports are provided automatically by the facilities management system printers.

Specifications also required that certain elevators have card readers that grant people after-hours access only to those floors for which their card has been programmed.

Residential Security Is Part of Program

The Stroh Companies security management program requires residential security for corporate executives and for those who will live in the multi-use Stroh River Place complex. Executive homes are monitored by officers at the security control center 24-hours-a-day for burglary, fire, smoke, temperatures, sump pump, and personal alarms.

The executive protection package has been designed to include dual signaling systems. Both digital signals, through existing telephone lines, and radio signals are received simultaneously from each protected premise. In addition, direct voice contact between the home and central station is included to aid in responding more accurately to any emergency call.

Residential units for tenants of River Place will have smoke detectors and infra-red sensors tied to the fire alarm system and the control center. Window breakage sensors and concealed door switches will be wired to fire security panels as well.

Additionally, a silent push button will be located in each bedroom of the River Place units along with an intercom that enables the resident to talk to officers at the security center. In the event of a break in, the resident could identify the location of the intruder in the apartment so police are totally aware of the situation when they arrive.

Systems Approach Works When Developing an Intelligent Building

A high level of security for people and property plays a major role in designating a building "intelligent." The systems approach to establishing a security management program is effective to develop a building or complex that is worthy of the term "intelligent."

To reduce crime and the fear of crime, those responsible for security must use a systems and team approach. It is not enough to put security officers on the premises and say, "Now we have crime prevention." It is not enough to install electronic equipment and say, "Now we have an intelligent building with security."

To best provide security for a corporation or industry, for a building or complex, those responsible for it must take into account the surrounding community. And, they must create a system that effectively teams trained and competent security officers with the sophisticated electronic security equipment available for today's intelligent buildings.

Chapter 6

Integrated Maintenance Management

William E. Hamm
Planned Maintenance Systems

The advent of the intelligent building and the consequent installation of monitoring and control devices opens the door to numerous opportunities for improving the delivery of maintenance services at lower cost. It has long been known that the measurement of certain environmental factors within a building envelope can tell us about the integrity of the envelope and, therefore, identify needed repairs such as roof tile replacement, caulking, replacement of moisture barriers, etc. Similarly, the measurement of vibration, temperature, and power consumption can tell us to lubricate, adjust or replace parts of machines. When these "condition" measurements are integrated into a program of maintenance, benefits can accrue that go well beyond the simple discovery of an abnormal condition signalling needed service. This chapter will explore the synergy resulting from the integration of "building intelligence" with the maintenance processes and its impact on the delivery of maintenance services.

While the use of environmental and condition monitoring devices has been common in some processing industries for some years now, their use in commercial, institutional, and general office buildings has been rare. The cost benefits of monitoring equipment condition in an environment where downtime has a great cost associated with it are plain to see. In contrast, the savings cost that can be achieved through the installation of monitoring devices in other environments is more difficult to estimate in advance of installing automated systems. Certainly the installation of the networks, equipment, and devices required for a monitoring system is not inexpensive. It is only when the benefits of such a system are expanded to serve not only maintenance but also security, safety, energy control and plant operations that the cost/benefit ratio clearly favors automation.

This is true because many of the sensors that support one activity also support another. The tradeoff decision becomes much easier when the monitoring and control systems are effectively integrated with the information systems of those activities that maintain and operate the plant through a building communications network.

Whether we examine how maintenance opportunities might impact the automation of a building or whether we look at the effect automation has on the delivery of maintenance services depends on whether we are justifying the automation or are taking the best advantage of a decision to automate. The decision to automate and the degree to which automation will take place will probably involve many considerations including cost, aesthetics, security, improved building services, and the delivery of occupant services.

It is easy to place a value on occupant services, fire/safety considerations, security, and energy consumption management. The value of improved maintenance is not as easy to pinpoint. For one thing much of the cost of <u>not</u> delivering maintenance services at the best possible level show up in other budget areas such as capital equipment, major renovation, and loss of productivity. However, some aspects are assessable such as those explified in Chapter 26 on maintenance management benefits at the British Columbia Housing Authority.

The case will be presented in this chapter that substantial savings can be realized by acquiring and using building intelligence to improve the delivery of maintenance services. This savings can be realized at a relatively low additional investment level because often devices installed for security, safety, energy control, and operations can provide inputs that stimulate smart maintenance decisions. Further, in areas where the collectors, processors, and networks are already established or justified, the incremental cost of adding sensors that provide intelligence to maintenance can be minimal.

THE MAINTENANCE ORGANIZATION

Before we examine the opportunities that building and machine intelligence can provide for maintenance, we must first examine what is meant by maintenance and how it is traditionally performed. From that discussion, we will proceed to how building "intelligence" derived from monitoring and control devices can improve the delivery of maintenance services.

Maintenance and operations are the sum of those services that are performed to ensure that the work place is a safe, comfortable, convenient work environment and that the machines used in the work place perform as intended. Operations are those services, such as are performed by boiler operators and security guards, that operate and control the buildings and the installed equipment. Maintenance, on the other hand, are those services that do actual work on the buildings and installed equipment.

The major objectives of most maintenance departments are to reduce failures; to respond to requests for service within a given period of time; to extend the useful life of the buildings and equipment; to ensure an attractive, safe work place; and to minimize downtime during periods of space or equipment use. Generally, an unspoken objective is to be invisible.

These objectives are generally pursued by employing a multiple-craft organization that performs the tasks according to some combination of both pre-planned scheduled work strategies and reactive strategies. Most often the department is organized around the functions performed, with the crafts grouped within a function. A typical maintenance department might be comprised of the following listed shops, each performing functions similar to those shown.

MAINTENANCE SERVICES

Shop	Function
Engineering	Project work such as renovation, building structural work, repainting, equipment installation.
HVAC	Preventive maintenance and repairs.
Boiler	Preventive maintenance (PM) and repairs.
Building maintenance	Building inspection, minor repairs, lighting program, PM on minor interior equipment such as water fountains and fan coil units.
Janitorial	Daily and project cleanings such as tile stripping and waxing.
Grounds	Grounds work such as landscape work, snow removal, area policing, grounds repair and restoration.
Vehicle	Preventive maintenance and repair of vehicles.
Service center	Receives and records requests for maintenance services; dispatches maintenance "on call" personnel; often distributes PM work orders and records or files work order completion data for all departments.

MAINTENANCE STRATEGIES

The strategies for accomplishing maintenance services can be broken into five procedural types of work: routine daily tasks such as daily cleaning; periodic work such as equipment preventive maintenance, landscaping, project cleaning, and building inspections; project work such as renovating an office; emergency repair work and service call response; and planned repairs.

Determining how much maintenance needs to be performed to satisfy the routine daily tasks is relatively simple. Varying the delivery of the services both in terms of their content and frequency will yield feedback in a very short period of time. Short-term observation is the key to discovering job effectiveness for these tasks.

The delivery of periodic maintenance provides a mixed bag of job effectiveness opportunities. In landscaping, for example, it is relatively easy to observe the effects of task content and frequency on the effectiveness of service. However, the same cannot be said for PM. In the case of mechanical and electrical PM, the effects of task content and frequency are not readily apparent except at the extreme ends. There is a significant spread between service intervals that are obviously too frequent and intervals which are obviously too infrequent. One way to find out "How much is enough?" is to create alternative ways of gathering the intelligence about the condition of the building components. Scheduling maintenance on the basis of conditions could eliminate excess PM.

The most obvious opportunities for saving costs in the area of project work lie in the management of the projects themselves and by eliminating the need for restorative projects when maintenance proves less costly than restoration. There is also a very real savings opportunity in the early identification of those restorative projects for which time is of the essence in preventing further damage. Minor failures have a way of becoming major projects when given time. Better yet, one would like to be able to predict the occurrence of a failure in time to make the repair before the failure occurs, thereby preventing the occurrence of downtime and the snowballing effect of failure.

There are several opportunities for reducing costs and improving services in the area of emergency repairs or unanticipated failures. One opportunity lies in acquiring the ability to predict a failure and correct it <u>before</u> it occurs. This is a capability made viable by the intelligent building. Effective PM can reduce the number of failures, but it must be carefully considered because it can be more costly than a failure.

The preplanning of diagnostics procedures, and repair and replacement of those parts most likely to fail, can measurably improve the delivery of emergency repair services, especially for components where the nature of failure is predictable.

The use of condition monitoring to trigger specific diagnostic or repair work orders as required, based on the intelligence received, provides very exciting prospects for dramatic improvement in the delivery of service for less cost. The measurement of the effectiveness of these opportunities is in the number of failures, the downtime statistics, and the average cost of repair. Unfortunately, this measurement cannot be accurately made until after a significant one-time investment has been expended on monitoring and preplanning.

MAINTENANCE PROCESSES

The means by which maintenance services are delivered vary, but most organizations employ a mix of planned and reactive processes.

- The maintenance department often performs preventive maintenance on equipment which requires lubrication, calibration or cleaning to function properly. These tasks are preplanned and are scheduled on fixed intervals. Most often a work order or task list is written which identifies the equipment to be worked on and the job to be done.

- The purpose of building inspections (performed on a periodic basis) is to discover needed repairs. The list of repairs is then scheduled for completion. Often, check lists are used for the inspection and work orders are issued for the needed repairs.

- Maintenance schedulers almost always plan project work in advance and may or may not use work orders. Most projects are for renovation, major repairs or special exhibits.

- Repairs required for machines or building components that affect comfort or safety are most often performed without pre-planning. Individuals or crews are dispatched by the Service Center, which receives and manages all requests for service from the building occupants.

Cleaning and grounds work is almost always performed on a fixed pre-planned schedule. The major exception is grounds work made necessary because of damage from natural catastrophes like tornados, lightning and heavy rain.

The delivery of maintenance services involves both static and iterative processes. Most preventive maintenance (PM) is planned once and is thereafer performed at fixed cyclical intervals. Generally, the actual work, required parts, and intervals between service are derived from manufacturer's recommended procedures. Manufacturer's procedures are mostly written for an assumed application in an assumed environment. Both assumptions can be wrong. A better source of PM procedures and requirements is engineering evaluation of the manufacturer's standards and the adaption of those standards to the in-house environment. The ideal set of PM procedures is one that keeps PM, repair, and replacement costs to a minimum while satisfying plan requirements relating to downtime.

PM is scheduled on the basis of calendar days between service, run time since last service, or on an observed condition of the component. Each of these methods of scheduling has its advantages and disadvantages, and the best method may be different for different components. For example, a job that can be scheduled in advance can generally be accomplished at less cost than if it were performed without advance notice. This argues for scheduling based on a calendar. However, the number of days between service is actually required. In determining when service is necessary, the passage of time is not as important as the time in use, the environment and the amount of work (load).

Cyclical scheduling can result in too much time between service one time and too little the next time.

Scheduling PM on the basis of run time is clearly more effective than the use of cyclical schedules in terms of reducing variation from the optimum point to perform the service. However, run time does not reflect the effect on a machine of environment nor work load factors and will therefore still vary from the optimum schedule. Further, run-time scheduling suffers some loss of productivity due to the scheduling problems associated with reactive vs proactive planning. Because scheduling on the basis of run time permits work scheduling to be done only a few weeks or days in advance of work performance, a system for handling a continuous resetting of priorities is required.

The major drawback to scheduling PM on the basis of calendar days or run time is that neither method really relates directly to a condition of the component. In both cases, maintenance can be early or late as compared to the optimum time. Setting aside considerations of scheduling personnel, the optimum time to perform PM is just before damage to the component occurs or just before there is a material increase in the amount of energy required for the component to function. The best known way to find this optimum point is to monitor operating characteristics of the component for changes that will indicate if an "off-normal" condition exists. The challenges in employing this method are, first, to discover those characteristics that will signal an off-normal condition; next, to develop a method of monitoring the characteristics; and, then, to develop a range of acceptable measurements that will yield reliable alarms (off-normal notices). Measurable machine characteristics include vibration, temperature, pressure, and power consumption. Characteristics of building spaces include temperature, humidity, air particle content, and electricity consumption.

The greatest disadvantage of using a condition-based measurement to initiate PM is that it is basically a reactive strategy which complicates the scheduling of downtime and personnel. Furthermore, it cannot eliminate the need for all cyclical maintenance -- such as start up, shut down, major overhauls, and other tasks requiring manual intervention -- because the installation of monitoring and control devices is not cost effective.

The optimum PM scheduling solution for each component may never be achievable. But if we examine each component in the light of what makes it fail, the consequences of failure, the clues to "off-normal" conditions, and the costs associated with PM, repair and replacement, then intelligent choices can be be made.

The delivery of corrective maintenance is often a highly iterative process. We often see the following series of steps:

Step 1. The need for repair is discovered and recorded.

Step 2. Maintenance dispatches an employee to define the problem and provide a fix if possible.

Step 3. The maintenance person returns having identified the actual problem. He collects the necessary tools/equipment and parts.

Step 4. The maintenance person returns to the location of the problem and attempts the repair.

Opportunity for improving this process lies in reducing the fault diagnosis step by installing monitoring devices that will not only report the failure but also identify possible causes.

These same devices can provide an earlier discovery of a condition requiring correction. Early discovery permits earlier corrective action and that in turn reduces the scope of corrective action by reducing or eliminating "consequent" damages.

HOW DOES MAINTENANCE AFFECT THE COST OF OWNERSHIP?

Now let us examine how both PM and repair can affect the costs of ownership and then identify those factors which influence the effectiveness of both PM and repair. We shall conclude with specific examples of how the intelligent building can contribute to the equation.

For the purposes of this discussion, the "annualized life cycle cost of ownership" ignores the impacts of the design and construction of the original facility and of any renovations and alterations.

Greatly simplified, the annualized cost of ownership is as follows:

$$A = \frac{C + R + U + PM + P}{L}$$

Where:

A = Annualized life cycle cost,

C = Construction cost plus taxes,

R = Repair costs,

U = Utility costs,

PM = Preventive maintenance costs,

P = Production loss or utilization loss,

L = Life of facility in years.

There are many more factors that affect life-cycle costs. Only those that are directly influenced by maintenance and, of course, the cost of construction are included in this discussion.

Inclusion of all the factors would make the examination of the impact of maintenance on ownership cost unnecessarily complex. Our intent is not to prove measurable impacts on costs but rather to demonstrate that the proactive management of maintenance can result in significant costs savings.

REPAIR COSTS - decrease as PM costs increase within a range of PM expenditures. This is generally true for any components for which PM tasks defer failure, such as lubricating points where friction exits. There are tasks that may be performed as PM but that affect only the correct functioning of a component and have no effect on failures.

Variance of these types of PM will affect user or occupant satisfaction but do not affect cost. With respect to those PM tasks designed to forestall failure or prolong useful life, the decision concerning how much is enough is an important factor in achieving the lowest overall cost of ownership consistent with objectives for machine use and downtime. However, overall lowest cost of ownership does not necessarily mean fewest repairs. This is especially true for low cost components whose failures do not interrupt occupancy or production.

REPAIR COSTS - increase as elapsed time between occurrence of the failure or off-normal condition and its repair increases.

For example, a motor that drives a fan loses its firm mounting and begins to vibrate. If adjusted and tightened right away, no damage may be done. However, if allowed to continue vibrating, the fan may strike its cage and damage both the fan and the cage. The fan may become wedged and prevent the shaft of the motor from turning resulting in the motor burning out. Or the vibration could cause an unequal weight distribution on the bearings that hold the motor shaft, causing them to wear and freeze.

In another example, the wind blows away a few roof tiles. Rain gets to the roof insulation causing it to pack and lose its insulation quality--which results in higher energy loss.

Water collects on the ceiling and spots it, necessitating painting, or worse, replacement of part of the ceiling.

As can be seen from the examples, the timeliness with which failure conditions (off-normal) can be corrected can affect the amount of damage that will result from an initial failure.

UTILITY COSTS - increase as the elapsed time between the occurrence of an off-normal condition (failure) and its repair increases. The loss of roofing tiles provides a heat leak and increases the energy costs to heat a space.

UTILITY COSTS - decrease with an increase in PM costs within a range of PM costs. A motor that is not well lubricated will require more energy than a motor that is well lubricated.

A fan pushing refrigerated air will consume more energy to move the same amount of air if the filter is dirty because of the increased resistance offered by the filter.

PM COSTS - vary directly with the extent of work specified for each interval of work and with the frequency of the performance of the PM interval. The cost impact of changing oil and filters every 4,000 miles instead of every 2,000 miles in vehicles used under normal driving conditions and in moderate weather is a savings of $300 for the average life of the automobile. For other than high-performance engines, modifying the schedule as stated above may have virtually no impact on effective life of the engine. On the other hand, changing the cycle for calibrating front-end alignment from every six weeks to every six months for automobiles that see a lot of travel over rough roads can reduce the average life of tires mounted on the front by one-half the expected life. More important, stretching the interval in this case could create a safety risk for the operators and occupants of the vehicle.

PRODUCTION OR UTILIZATION LOSS COSTS - increase as the amount of time between the occurrence of a failure that affects production or occupancy and its repair increases. In many of today's modern buildings, occupants are not able to open the windows. When the air conditioning goes out on a warm day, so do the people. Another example is inadequate lighting which has been shown to cause reduced productivity.

PRODUCTION OR UTILIZATION LOSS COSTS - decrease as the investment in PM for work designed to prevent failures affecting production machines or space occupancy increases, within a range of PM costs.

COMPONENT LIFE - extends as the investment in PM increases within a range of PM costs for work designed to prevent degradation. Caulking and painting wooden shingles and fixtures prevents moisture from causing rot. Lubricating moving parts slows friction wear and extends life.

COMPONENT LIFE - decreases as the time between the occurrence of an off-normal condition and its repair increases. A frozen bearing on an operating motor is likely to burn out the motor if not discovered and corrected immediately. An out-of-alignment wheel can destroy a tire in a very few miles.

The above discussion of factors affecting life-cycle costs provides several major challenges to management:

- Which component should receive PM?

- What tasks should be performed as PM?

- How often should PM be performed?

- How can failures be detected and corrected in the earliest stages?

The last two of these challenges can be directly assisted by the intelligent building.

It is said that the performance of PM accomplishes three objectives: 1) to make a machine run correctly or more efficiently, 2) to prevent a failure from occurring, and 3) to discover an off-normal condition which requires corrective action.

All three of the cited objectives are targeted to correct or prevent an off-normal condition. Yet, as we have seen, most PM is scheduled on the basis of calendar week or run time. Neither of these methods reflects the actual condition of a component. Sometimes PM comes late, and sometimes it is performed early using these scheduling techniques. Ideally, we would schedule PM just sufficiently prior to the occurrence of off-normal condition to allow for cost effective personnel scheduling--perhaps one week's notice.

The monitoring and control systems discussed in this book are ideally suited to provide intelligence relating to the condition of machines and structures. We can integrate that intelligence into a condition-based maintenance program. The observation of condition measurements, and particularly of the changes in condition as they relate to failures and less severe off-normal conditions, can lead to a program of predictive maintenance.

Predictive maintenance is the ability to schedule and perform corrective or preventive maintenance just prior to the occurrence of off-normal conditions. Sometimes more than one measurement or condition is necessary before a prediction can be accurately made.

But, in most building system's environments, an off-normal condition can be detected for most plant components in sufficient time before failure to permit scheduling of the corrective action.

Whether or not the installation of monitoring and control devices is justified depends on many factors for each of the many components of the physical plant. One major consideration is the frequency of visits to the component to perform work which is not based on component condition or for work whose schedule is dicatated by considerations other than equipment condition. If the visits required by considerations other than condition are frequent enough to prevent or discover and correct an off-normal condition, then condition-based maintenance may not be worth the investment for that component.

Another important consideration in deciding whether or not to install monitoring devices is the extent to which energy costs can be saved by timely maintenance. Many machines can operate for long periods of time in off-normal conditions without failing.

Yet, these machines will consume much more energy than if they had been maintained in good operating condition. <u>The Department of Energy estimates energy savings resulting from scheduled maintenance of between 5% and 15% of the total energy bill.</u>

Any highly efficient maintenance program is likely to contain cyclical PM, run-time PM and repair, condition-based PM and repair, and preplanned repair and diagnostic procedures.

Cyclical maintenance is going to be used wherever scheduling of maintenance is dictated by other considerations. One example would be the purging of steam lines during Christmas vacation at a college. Even though energy costs might be reduced by purging sooner or labor costs reduced by purging one month later, in this instance the practical considerations related to the effect of downtime overweigh cost effectiveness.

Start-up operations and shut-down operations are most often associated with a date and will therefore be scheduled on a cyclical basis.

Inspections which are scheduled to observe conditions that cannot be economically monitored by devices are going to be scheduled on a calendar basis. Examples are inspections for aesthetic conditions and minor structural problems such as loose floor treads or tiles.

Some off-normal conditions are so predictable that scheduling based on the calendar cannot be improved upon. An example would be the drawing of water from steam radiators used in a domestic setting. While the predictable interval may vary widely from site to site, the condition is generally predictable within a given environment.

Scheduling on the basis of run time is going to be done whenever it is more cost effective than on the basis of condition, and run time can be monitored. Run time is an excellent basis for scheduling brake and alignment service on vehicles.

Another use of run time is the scheduling of lubrication on mechanical parts.

Scheduling maintenance based on condition is primarily dependent on the cost of installing devices and monitoring versus potential cost savings. Here are some potential uses of condition-based maintenances:

- Electrical control circuits can be monitored for connectivity and leaks. These measurements provide ample intelligence to schedule both PM and repair based on condition.

- Monitoring electric power consumption can be used to schedule lubrication, filter changes, cleaning and adjustment exactly when the service is required.

- Monitoring fluid viscosity can be used to initiate timely changes of oils and lubricants.

- Sensing the amount of light can trigger work orders to clean light reflectors and/or replace bulbs.

- Monitoring gases can be used to initiate maintenance on combustion components.

- Monitoring vibration can be used to generate work orders to calibrate, adjust, and tighten mechanical parts.

- Monitoring pressure can be used to generate work orders to change or clean filters, purge systems, and clean or replace safety valves.

- Monitoring temperatures can be used to generate work orders for cleaning heat exchangers, lubricating mechanical parts, adjusting burners, and cleaning electrical circuits.

Perhaps the best way to use condition-based maintenance is to discover off-normal conditions precedent to failure in order to generate replacements or repairs prior to actual failures, especially when a failure can cause additional damage or when downtime is disruptive or costly.

The use of condition as a basis for scheduling maintenance can end up more costly than it need be if you do not view each visit to a machine as an opportunity to perform other work which might be done in the near future. Often a substantial part of the cost of maintenance labor is getting there and back. For example, a process that had you respond only to an off-normal alarm could have you make one trip this week to change a filter and another next week to lubricate a bearing. To these off-normal alarms, let us add a scheduled safety inspection for the second week. When you compare this three-visit scenario to the standard scheduled PM in which all three tasks are performed every quarter with one visit, you can see that you could soon lose costs saved by scheduling the lubricating and filter change on the basis of condition.

Because scheduling is made more difficult if much of PM is condition based, an even better way to use condition-based maintenance is in conjunction with either cyclical or run time scheduling. By combining two methods, one can stretch the intervals between scheduled service to the point where the monitoring devices will only occasionally detect an off-normal condition. In this manner, the advantages of scheduling are achieved most of the time, maintenance is sufficient to prevent premature failure, and very little more preventive maintenance is performed than is necessary.

Finally, we must take one additional factor into consideration when employing condition-based maintenance. That is that condition-based maintenance may extend the life of a component beyond the point when it would be more economical to replace than to maintain. In an environment where condition is not monitored, parts and machines often fail before an off-normal condition is observed and as a result will be replaced more often than in a monitored environment. To overcome this problem, it is necessary to build in a parts or machine-replacement criterion.

Chapter 7

Integrated Lighting Management

David Peterson
General Electric

"Smart Doesn't Necessarily Mean Better"

Recent press on intelligent buildings reminds us of the many experiences we've had with people who were brilliant but had no common sense. At first we tend to be awestruck with their mental prowess; but after a while we begin to wonder how they ever manage to live in the real world. Buildings that bristle with "smarts" but don't really understand or respond sensibly to the needs of the people who live in and operate them, won't do well in the real world either.

The real challenge in designing an intelligent building isn't a technical one of computers, bits, and bytes; it's a practical one. . . "How do I increase the value of this building by making it respond sensibly to the needs of the occupants and operations people?" This is an applications problem; one in which the people who understand the needs of the occupants and the practical considerations in managing a facility have the most critical input.

This is particularly apparent when it comes to lighting and lighting control. In the real world, people don't think very much about lighting; they simply assume that it's going to be there when they need it. In fact, automation of the lighting may be viewed as a negative: "In the past it was always ON, now 'Big Brother' is going to decide where and when I can work." Designers who fail to give individual occupants the ultimate control of the lighting in their spaces have to be prepared for negative, emotional responses. The lighting automation system may be viewed as just one more attempt to restrict personal freedom.

The tragedy in this situation is that the frustration and confrontation could have been avoided had the designer specified the lighting controls around the needs of the occupant. The technology and products exist which can not only meet the productivity needs of the occupants but also can offer other salable features and save energy in the process. This chapter provides an overview of the needs which a lighting control system should address in today's office building and then compares how well some of the different approaches commonly employed fare against these needs.

Lighting control needs can be defined from both an occupant and an operations perspective.

The objective of any lighting control system is to provide:

- the right amount of light,
- where needed,
- when needed.

This objective, however, needs to be viewed from two perspectives: that of the occupant and that of the building manager. At the same time, controls must be viewed in the context of the lighting system intended for the space.

THE OCCUPANT'S PERSPECTIVE

The amount of light needed (or wanted) within a particular area is not static. It is a function of the nature of the task being performed and the age of the worker. For example, for typical reading tasks, the IES (Illuminating Engineering Society of North America) recommends lighting levels of 50-100 footcandles for workers under 40 years old and 75-150 for those over 40. On the other hand, reading a visual display terminal requires only 5-10 footcandles. Furthermore, cleaning crews need only 15-30 footcandles.

The important point is that the amount of light required is not a strict number. It varies by space usage and individual. Furthermore, the introduction of video display terminals (VDT) within the office environment presents a major crinkle in lighting system design. For those individuals working exclusively on VDT's, a very low general illumination is desirable. However, if the occupant mixes VDT usage and general reading tasks, the lighting system needs to accommodate both. With these points in mind, it is easy to understand why some leading firms are leaning towards greater flexibility in the lighting design and allowing occupants to control the lighting within their areas.

The _where_ and _when_ are more straightforward than the _amount_ of lighting. The basic position of an occupant is that he should have adequate lighting whenever he is in the building. The implication is that not only should his individual work area remain lit if he wishes to stay late, but he should also have lighting in those areas which he will tend to use (the copy machine room, restrooms, hallways, and elevator lobby). This represents a minimum requirement for any lighting control system to be occupant sensitive...it must provide a simple means for the occupant to override the automation for his space, and it must provide for safe access and egress to his area when he uses the building after hours.

THE BUILDING MANAGER'S PERSPECTIVE

The building manager approaches lighting control from a different perspective. He is more focused on issues such as:

1. Energy cost reduction,
2. Personnel management,
3. Security,
4. Space usage adaptability.

Energy Cost Reduction

Lighting typically represents 30-50% of the buildings electrical power consumption or approximately $0.60/sq. ft./year.

Since lighting controls can reduce this expense significantly, they are an important ingredient in the energy management programs for buildings.

A lighting control system may use several different strategies for reducing the energy consumption:

1. Occupancy-based control of lighting levels.

This is normally the most significant source of energy savings. Typically the lighting levels are scheduled based on occupany -- full lighting for normal occupancy, reduced levels for the cleaning crew. This is also the function that can lead to numerous occupant complaints if the system doesn't allow the individual occupant to exempt his area when he wishes to work afterhours.

2. Daylighting control.

Reducing the "artificial" lighting when natural daylighting is available has a strong emotional appeal. To take full advantage of daylighting, however, the building should incorporate several design features. These include proper siting and orientation, provisions for glare control, and design of windows and light shelves to enhance daylighting penetration. The building should also be wired to provide lighting zones 10-15 ft. deep running parallel to the windows. For these reasons, daylighting control is normally restricted to new buildings designed for its use.

3. Tuning.

Tuning provides the ability to vary lighting levels to reflect the actual amount needed for the task and age of the occupants. The tuning may occur on a fixture-by-fixture basis or for a large zone. For example, assume that the general overhead lighting has been designed to provide 70fc (footcandles) of lighting throughout a floor. Individual fixture tuning may allow lighting in walkways to be reduced to 20-30fc without affecting the lighting in the critical task areas. Large zone tuning, on the other hand, might be used to provide "task-ambient" lighting for a department using VDT's extensively. In this case, the overhead general lighting would be reduced to approximately 30fc and each occupant would be provided a task light (desk light) to provide the 70fc level on the reading task only.

Daylighting and tuning provide energy savings by reducing the lighting power requirements (usually measured in watts/sq. ft.) during normal work hours. They also have a positive impact on building electrical demand charges. Scheduling, on the other hand, provides savings by reducing the runtime. Since the afterhours reductions do not normally coincide with peak loading periods, there is no reduction in electrical demand.

Personnel Management

The manager's immediate concern is that the lighting control system not require incremental people or special skills.

On the positive side, the lighting control system may provide valuable information. For instance, if all lighting overrides can be centrally monitored, the operations manager can monitor the cleaning crew's progress through the building or track the security guard making his rounds.

Security

Central monitoring of the lighting also makes it possible to alert security personnel. For instance, the card access system might provide a command to the lighting control system to turn the lighting ON for an occupant entering the building after hours.

If the occupant turns ON other lighting zones, the security personnel might be alerted to make a check of the area.

Fig. 7-1 Lighting Control System Elements

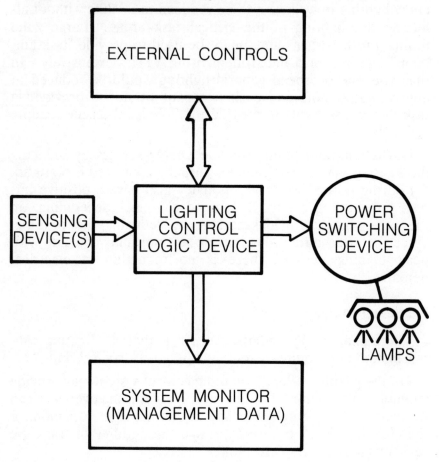

In addition to monitoring space usage, the lighting control system can provide a valuable extension to the security system. As a guard makes his rounds through the building, the lighting controls can provide greater safety by automatically lighting the space he is about to enter while turning OFF the area he has just exited. Also, any security alarm could automatically turn ON the lighting in the area affected.

Space Usage Adaptability

The ideal light control system should not require additional work and expense when spaces are rearranged. In particular, the building manager would like to avoid controls which require rewiring of fixtures or sensors whenever the floor plan for a space is changed.

Incremental Income

A system which would allow the facilities manager to increase revenues, not just reduce expenses, would be ideal. This might be accomplished by controls for which the occupant pays extra rental charges or by controls which provide a simple means for charging for overtime usage.

ELEMENTS COMMON TO ANY LIGHTING CONTROL SYSTEM

This sketch (see Figure 7-1) helps illustrate concepts common to any lighting control system. The "Control Logic" block represents the "brains," or thinking capacity, of the system. The "brain" receives information from either "Sensing Devices," such as switches or photocells, and from "External Controls," such as the card access system or the security system. The "brain" processes these inputs and activates the power switching device to the appropriate level. These process decisions may be reported back to the "External Controls" to confirm that the inputs have been received and acted on. The actions and status of the control logic system may also be reported to a system monitor for better management and

Fig. 7-2 Power Switching Device

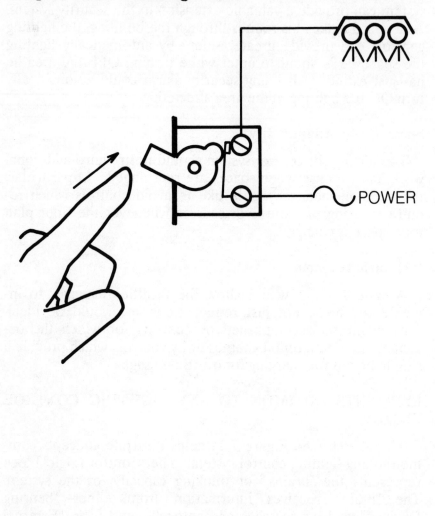

maintenance. The flow of information from one block to another is a function of the system's communications capabilities.

Alternate System Configurations

The concept diagram above helps clarify some of the key functional differences between systems which are either on the market today or represent logical extensions of these systems.

Standard line voltage switching: The common wallswitch (see Figure 7-2) is the simplest system available, but it is also limited. Using our control model, the toggle is the sensing device...it senses occupancy by requiring an action. The brain is nothing more than a cam mechanism that opens and closes a set of power contacts (the power switching device.) The common switch doesn't receive information from external controls nor does it provide management data.

People presence detectors: These devices (see Figure 7-3) typically use ultrasonic disturbances or infrared radiation to sense the presence of an occupant within an office area (normally 150-200 sq. ft.). The on-board "brain" then sends a signal to close a relay which is connected to the lighting circuit. If the brain doesn't receive an "occupied" signal from the sensor for a period of time, it turns the relay OFF. In essence then, this is a switch with improved sensing and logic capabilities. Like a switch, however, it is weak in receiving signals from external control devices, such as a photocell, and in providing management data.

Programmable Lighting Control System

This approach (see Figure 7-4) is based on a central computer which communicates with intelligent field panels over a single twisted pair dataline. The central computer initiates all actions to the lighting relays based on:

1. Its internal operating schedule;

Fig. 7-3 People Presence Detector

ALTERNATE 2-

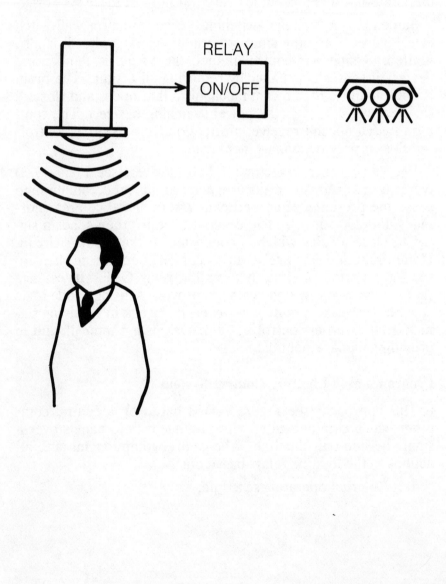

2. Signals from photosensors or occupant switches connected to the field panels;

3. Occupant overrides via their existing telephones; and

4. Input commands from external controls, such as the card access system or security system.

Since these commands all originate in the central controller, system activity can be readily monitored and processed to provide management data such as the overtime usage billing report.

Referring again to our general block diagram, (see Figure 7-1), you can see that this system is based on a strong communications capability which leads to better integration with external control systems and provides valuable management data. Software focused on lighting control also provides a more occupant-friendly approach to lighting automation than would typically be provided by putting lighting contactors on the building automation system. For example, the floor plan sketches (see Figure 7-5) illustrate how the lighting for each zone might be scheduled according to normal occupancy.

In actual practice, however, individual occupants don't follow a rigid schedule. Rather than simply turn OFF the lights and allow them to grope for the override switch, this system would provide a warning flick of the lighting. Occupants wishing to stay late would then have five minutes to enter a priority override for their area with either a wall switch or their phone. These overridden areas would then be exempted from the scheduled OFF sweep (see Figure 7-6). They would also be protected from the actions of the cleaning crew switches. These have been programmed to provide partial lighting in the area being cleaned and to turn OFF the area exited when a new area is entered.

Electronic Dimming Systems

So far we have progressed from switches to "automatic" switches to a centralized switching system. In each case,

Fig. 7-4 Lighting Control Focused Software
TYPICAL SYSTEM CONFIGURATION

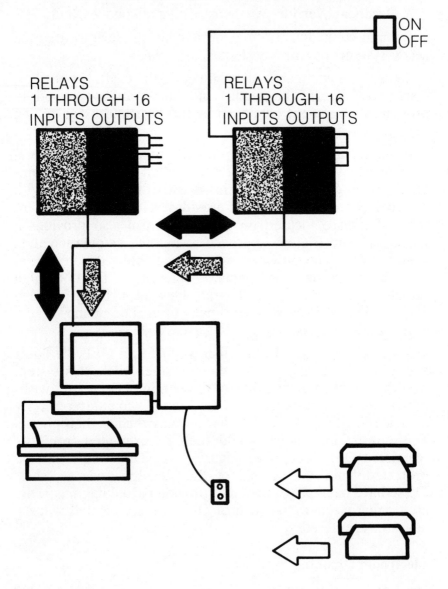

however, the power switching device was a contact closure providing ON/OFF control of a lighting circuit. Split wiring the lighting fixtures provides a low cost method for obtaining different lighting levels in a zone. As shown in the sketch (see Figure 7-7) of a three lamp fixture, selective switching allows OFF, 1/3, 2/3, and ON. (Note: The two relays would typically control 15-20 such fixtures in a zone of 1000-1500 sq. ft.). This ability to provide different levels greatly increases the savings potential when compared to a simple ON/OFF control approach. Daylighting strategies, in particular, are practically impossible to implement without multiple level capability. Multiple level switching may also be applied to individual offices to allow occupants to select their own lighting levels.

ON/OFF switching does have drawbacks, however. When implementing a daylighting strategy, for example, the occupants notice the sudden change in lighting output when the fixture is switched from one level to the next. Furthermore, some people object to the appearance of a fixture with some of its lamps switched off. Finally, switching does not provide the resolution required to tune the lighting levels to that required for different tasks.

Substituting a dimming module (see Figure 7-8) as the power switching element provides these added features. A dimming module has been substituted for two relays in the programmable lighting control system simply to illustrate the concept. All of the features of that system would be maintained, but now the lighting output of the circuit could be varied continuously.

Dimmable Solid-State Ballasts

The above dimming approach assumed the use of standard ballasts in the lighting fixtures. Dimmable solid-state ballasts are now available which provide improved efficiency coupled with dimming capability. The dimming function, however, allows a fixture to be dimmed automatically as part of a group and to have its maximum output individually adjusted.

Fig. 7-5 Occupancy Based Scheduling

Schedule ON by Area (or Occupants Override ON)

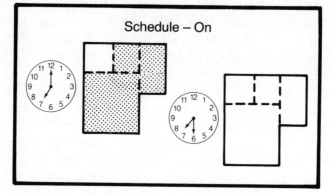

Schedule Warning "Flick" Prior to OFF Schedule to Avoid
Putting Occupants in the Dark*

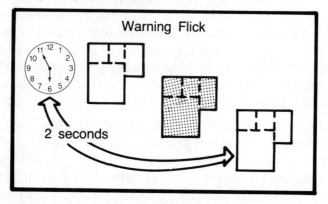

Occupants Enter Protected (Priority) Override for Their Areas

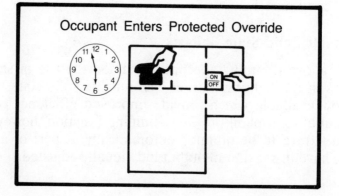

Fig. 7-6 Exempted Areas

Occupants Exempted From OFF Sweep at 6:00 pm

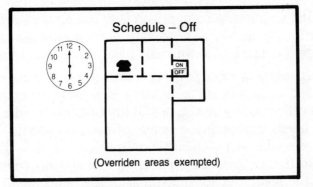

Same process (warning, override, sweep) repeated throughout the night.

2. Reduced Lighting for Cleaning

Alternative #1 . . . Schedule Reduced Levels With Occupant Override

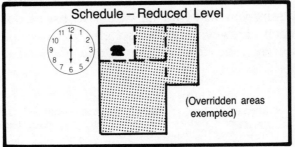

Alternative #2 . . . Interlocked Cleaning Crew Switches. Turning ON the Area Just Entered Automatically turns OFF the Remainder of the Floor

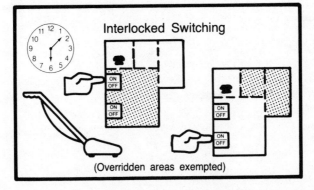

COMPARING SYSTEMS AGAINST YOUR NEEDS

The first step in any comparison is to understand your building's particular needs. These vary not only by building type and usage but also by the way individual areas are configured.

In particular, individual office areas are very different than landscaped offices from a lighting control perspective. Individual offices open up the possibility of adjusting the general lighting levels to match the needs of each occupant. In open office areas, this is practically impossible since a single overhead fixture may provide lighting to several occupants...who decides the level? Also, individual offices lend themselves to occupancy sensing devices more readily than large open spaces.

Given the assumption that you've done your homework on the building's usage, how do you evaluate different approaches to lighting control? If you read the manufacturer's literature on any system, you will find an extensive list of positive features for each. However, no two systems will contain the same set of features. The key problem is one of "normalizing" the

Table 7-1

	Occupant Sensitivity	Light Level Selection	Energy Savings Potential	Management Data	Integration Capability	Space Adaptability	Cost
Standard Wall Switches (Individual Offices)	Good	Fair	Fair	No	No	Poor	Medium
Contractor Control Via Building Automation System	Poor	Poor	Poor	Yes	Yes	Good	Low
Programmable Lighting Control - Relay Based	Excellent	Far	Good	Yes	Yes	Good	Medium
Programmable Lighting Control - Dimming Based	Excellent	Excellent	Excellent	Yes	Yes	Good	High
Occupancy Sensor	Fair	Poor	Good	No	No	Poor	High

features so that they can be readily compared. Table 7-1 shows a simple comparison of different systems according to how well each meets the needs of both the occupant and the building management. Of course, it would be a simple procedure to provide a weighting on each function from your building's perspective and then compare the weighted performance of each against its cost.

One word of caution; the relative performance rating of each system is often a judgement call. Individual manufacturers for a single system may vary sharply in system performance and cost. The objective here is to show a process...not pass final judgement or provide hard pricing guidelines. The reasoning that went into the relative rating for each function shown in Table 7-1 is summarized below.

Reasoning

1. Occupancy sensitivity:

From an <u>occupant's</u> perspective, individual wall switches work just fine. Contactors on building automation systems can be a real negative, however, since they don't normally allow the occupant to override for afterhours usage. Occupancy sensors may provide lighting whenever the occupant is in the office, but the "space-age" effect can be eerie. Programmable lighting control caters to the occupant. When he enters the building after hours, his area can be lit in anticipation of his arrival with a single phone call. Similarly, when staying late, both his office and related work space can be kept ON with a phone call or switch override.

2. Occupant level selection:

This is a function impacted both by control systems capability and the floor layout. Individual office layouts with manual switching of split-wired fixtures or several lighting sources within the space give most occupants the degree of control they need. This can be done in conjunction with controls providing automation of a zone. The dimmable solid state bal-

Fig. 7-7 Splitwiring For Different Levels

ALTERNATE 3
PROGRAMMABLE LIGHTING CONTROL

STEPPED LIGHTING LEVELS

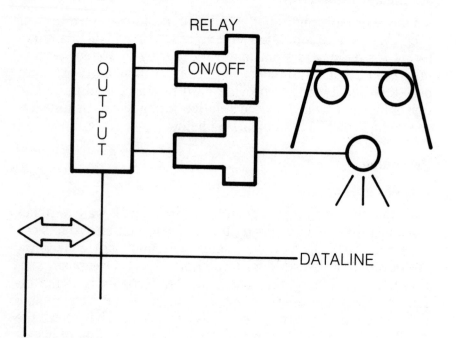

last approach provides an even better method for allowing the occupant to adjust the overhead lighting.

3. Energy savings potential:

Perhaps the only surprise here is in the "good" rating for switches in individual offices and the "poor" rating for the same devices used to control a zone. In practice what happens is that when more than one person is in a zone, the first in turns it ON and the last out never looks back.

4. Management data:

This term reflects system monitoring, analysis, and reporting capabilities. These require a communications capability not normally inherent in switches or occupancy sensors.

5. Integration capability:

Again, to integrate lighting with other building automation functions, the lighting control system must have good communications capability or be part of the building automation system.

6. Space usage adaptability:

The important point here is that devices which are physically linked to the occupant's walls or ceiling pose problems when it's time to rearrange a space.

7. Costs:

These are for illustration purposes only. The solid state ballast costs, in particular, represent a combination of functions not presently available in the market. Perhaps the biggest surprise is that individual office switches aren't cheap. These typically cost $50-$100 (two hours installation time) in new construction which leads to the cost of $0.50/sq. ft. Occupancy sensors may reduce this installation labor (one firm reports a one hour installation time as typical) but the added hardware content still means a relatively high total cost. Again, both switches and sensors incur added cost for office rearrangements. The other systems gain their economic advantages by controlling full lighting circuits (they are lighting panel-based instead of individual office-based).

THE FINAL ANALYSIS

In the final analysis, the lighting control decision boils down to a question of who gets the benefit, and who pays. Common sense says that whatever system you choose, make sure that you don't turn off either the owner or the occupant. The owner may be paying for the design up front, but the occupant eventually pays for everything.

Fig. 7-8   DIMMING

ALTERNATE 4
PROGRAMMABLE LIGHTING CONTROL

DIMMING

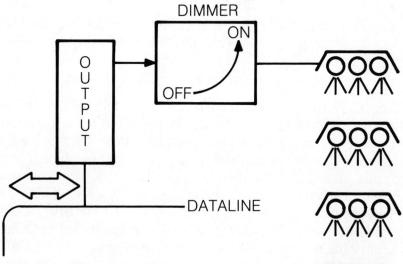

Chapter 8

Integrated Telecommunications Network Management

James H. Morgan
J.H. Morgan Consultants

This chapter covers a broad range of intelligent building subjects from a telecommunications management perspective. It is oriented to non-technical, management-type persons. Everyone involved in intelligent buildings must recognize that modern, sophisticated telecommunications has rapidly risen in importance to become a prime area which must be considered "up front" in building planning. In this regard, telecommunications now joins computers, environmental control, security, fire, and other traditional areas.

It is stressed here that good, solid telecommunications management is the secret to success in harnessing the power of telecommunications technology for the intelligent building.

INTRODUCTION

Telecommunications has become one of the fastest moving areas of activity, and will continue so for many years into the future. Some call the 1980's the "Decade of Telecommunications." Telecommunications has risen rapidly to now rival computers in rate of growth. In fact, these two areas of telecommunications and computers are heavily merging due to their interdependence.

Telecommunications has recently (and finally) gained upper management's attention. Telecommunications is now seen as a true resource, rather than a "necessary evil" which one must live with. As a resource, telecommunications is capable of producing income directly and also making money indirectly by increasing staff efficiencies, by delivering valuable information in a timely fashion, and by other means. Telecommunications is now recognized as a good area to invest in.

All of this advice applies directly to the intelligent building. There is money to be made and other advantages to be achieved by people paying attention to telecommunications -- building owners, developers, managers, tenants, architects, engineers, realtors, and consultants.

This chapter covers a broad range of intelligent building subjects from a telecommunications management perspective. Some subjects briefly discussed here are covered in more detail in other chapters.

MAJOR TRENDS IN TELECOMMUNICATIONS

The major trends in telecommunications, which must be recognized and acted upon by all parties involved in the intelligent building, are as follows:

- Control of telecommunications by yourself.

    You no longer can rely on any one organization to "take care of you."

    Telecommunications as a base for the oncoming Information Society.

    Massive amounts of information must be properly distributed.

- Move to "All-Digital World."

    The move away from the earlier "analog" world to "digital" offers many advantages.

- Integration of all forms of information/intelligence.

    The common denominator of "digital" allows telecommunications facilities to handle and integrate all forms of building intelligence.

The All-Digital World

A basic understanding of the word, "digital," is useful for those involved in the intelligent building, because the word is seen everywhere. It pertains to all types of building systems, not just telecommunications.

The basic concept in digital systems is that all intelligence/information is reduced to numbers. The Latin "digitus" means "finger," which we all used for counting as children.

Certainly the computer is naturally number-oriented. But now the human voice is converted to a series of numbers. This digitizing of voice was the breakthrough which is driving us to the all-digital world. After all, voice communication is still 90% of all communication.

This voice digitizing is transparent to the user, that is, not perceived during a phone conversation. What happens is that voice is converted to a series of numbers (coded) by the telephone equipment at the speaker's end, switched and transmitted in this digital form, and then reconverted to the original voice form (decoded) just before reaching the listener's ear. This is nothing new. It started in 1962. What is new is that semiconductor, microelectronics technology has made this coder/decoder process very inexpensive in the early 1980's.

Why go to all this bother? There are several reasons. It is much easier and cheaper to communicate information in digital (numerical) form. The accuracy, quality and reliability is much higher. Once digitized, voice information can be integrated, together with all other forms of digital information, so that telecommunications systems can be shared. Most buildings' people are aware that all building systems such as environmental control, security, fire, and others are moving heavily to digital. Certainly the relative newcomer, office automation, started right off with digital.

Just to complete this simple, non-technical explanation of the key word, "digital," it should be mentioned that the other, competing technology is the older "analog" world. Basically, an analog signal is wavy. Voice is the most common example, where the electrical voice signal is "analogous" to the wavy voice energy from the vocal chords. Another analog example is a voltage level from a transducer, which varies up and down (wavy) according to a room's temperature (or humidity, etc.).

The building people's payoff with the all-digital world comes from the sharing of digital telecommunications systems; namely, the sharing of building switching (e.g., PABX) and transmission equipment, including building wiring systems. Access to the outside world, which is going heavily digital, is simpler.

Integrated Voice/Data/etc. Telecommunications

A popular buzz-phrase is "integrated voice/data." It is like the buzz-word "digital" in that one must use the term in any telecommunications discussion if one wishes to appear current.

Integrated voice/data is actually a convenient contraction for integrated voice/data/video/message/word processing/facsimile/personal computers/environmental control/security/etc. The original integration of just voice and data, which began in the 1960's, has continued to expand as new, digital-based technology has developed. The early 1980's saw a sudden surge. Digitizing of video and facsimile became practical. The digital microprocessor boom produced word processors and personal computers.

And very important to buildings' people was the addition of traditionally non-telecommunications systems such as environmental control and security. Their move to digital designs was a big factor. In some instances these building functions were designed right into new telecommunications products, such as digital PABX's, by bold vendors. In some instances, interfaces were developed to join these building functions to existing telecommunications systems.

A driving force for integration was the concern that a building would have over half-a-dozen layers of wiring. Presently, there actually are some buildings with four layers of wiring, e.g., voice, data, word processing, and environmental control.

The term "telecommunications" is used in this chapter in its broadest sense, i.e., covering integrated voice/data/etc. Earlier, some people equated telecommunications with "telephones" only, which of course is now outdated.

The term "network" is often used in conjunction with "telecommunications," because in virtually all telecommunications applications network is involved, i.e., a system of lines interlacing like the fabric of a net. This, plus the important trend to integration, all needs managing, hence this chapter's title, "Integrated Telecommunications Network Management."

TELECOMMUNICATIONS MANAGEMENT

The glamor of telecommunications technology (microcircuits, computerizing, satellites, microwave, fiber optics, etc.) has unfortunately caught everyone's eye to the detriment of proper telecommunications management. Certainly the technology is a major driving force, but it will go nowhere for users such as buildings' people if this force is not managed/controlled properly. Good, solid telecommunications management is the secret to success in harnessing the power of telecommunications for the intelligent building.

Frankly, users are just not experienced in managing their own telecommunications. Up until ATT/Bell divestiture in 1984, "Ma Bell" would "take care of you." They installed and maintained the building wiring, PABX switch, outside transmission, and so forth.

Prior to that, as a result of the 1968 Carterphone Decision, whereby users could attach non-Bell "interconnect" telecommunications equipment to the Bell system, a number of the more adventurous users (thousands) stepped out and took more control of their telecommunications. These users installed their own PABX's, private networks, and so forth. Many were tenants in buildings. They learned the hard way.

Some were very successful, some had middling success, and some did poorly.

It is from these users that developers of intelligent buildings can learn. There is no need to re-invent the wheel. Intelligent building projects will have enough of their own unique telecommunications problems.

Again, as stated earlier, the secret to success will be proper telecommunications management. It turns out that the majority of telecommunications problems are not really technology-based, although so many at first glance appear to be. Rather, the majority are management-based problems. If the reader is not ready to accept this fact and plan strategy accordingly, he/she would best drop out of the intelligent building arena now.

Keys To Success

Some of the more important keys to success in managing telecommunications for intelligent buildings are as follows:

1. Provide your own, dedicated telecommunications staff, to control your telecommunications.

2. Make telecommunications a prime area of interest up front, not as an afterthought.

3. Assign telecommunications equal importance with data processing/management information systems (DP/MIS); and in fact, join the two under an Information Management umbrella.

4. Provide sufficient, quality floor space for telecommunications, including a network control center, plus possibly secondary telecommunications areas.

5. Provide suffient, readily accessible ducts/conduits for building wiring.

6. Accept the need and cost for continual, strong efforts in planning, training and documentation, three areas which traditionally lose out to day-to-day operational exigencies.

7. Design flexibility into telecommunications systems to better accommodate the rather uncertain future, even though cost is somewhat higher.

8. Plan telecommunications strategy to best meet the needs and desires of tenants.

These important issues will now be discussed in more detail.

1. Dedicated Telecommunications Staff

Telecommunications is too crucial a function to hand off to others. People involved in the intelligent building will be tempted to say they are building experts, not telecommunications people; they know too little about telecommunications; they should leave telecommunications to the telecommunications experts. These people must push temptation aside, bite the bullet, and openly accept telecommunications as a full-fledged function they must manage, just as they have in the past accepted environmental control, security, fire, elevators, computers, and other building functions.

This means hiring dedicated telecommunications staff. Just how big a staff depends on many diverse factors. The telecommunications staff can range from one to dozens, depending on: size of building; type of tenants sought; degree of building intelligence desired; amount of telecommunications assistance farmed out to service organizations; etc.

Regarding the latter, intelligent building management could decide to use minimal telecommunications staff and have most telecommunications handled by vendors, common carriers, third parties, etc. -- or handle all telecommunications themselves. Either telecommunications management approach can work. The main point is to establish and maintain telecommunications control. Never accept an outsider's "trust me" approach, even if it appears safe and even if telecommunications seems a bother to some bricks-and-mortar people. Recognize "you are on your own."

Telecommunications control by internal staff would cover all phases of telecommunications, namely, planning, specifying, selecting vendors, contracting, installing, and testing/accepting, followed by the continuous operations and maintenance.

2. Place Telecommunications Up Front; Do Not Make It An Afterthought

The building architect considers many functions "up front," such as flow of people, air, heat, air conditioning, lighting, water, sewage, electricity, telephones, and so forth. These functions have been traditional factors in building design. Then computers (DP/MIS) were added, -- initially as an add-on and later as a prime function.

Now, modern telecommunications must be added up front, which goes well beyond the relatively simple task of running telephone wire pairs into rooms. The relatively "dull subject" of building wiring has now become a challenging, sophisticated subject. The simple telephone is being replaced by the microprocessor-based workstation. The distribution of high-speed data is becoming common. Teleconferencing must be provided. Cables carrying integrated signals with all-eggs-in-one-basket must be well protected and made redundant. Local area networks must be accommodated. A wide range of telecommunications suppliers is available. All transmission media must be planned for, namely, twisted pair, coax cable, fiber optics, infrared, microwave radio, and satellites. Building wiring and the broader subject of telecommunications can no longer be treated casually.

Older buildings are being retrofitted for today's telecommunications, somewhat like an afterthought, but compromises must be made. Good architects of new buildings have now placed telecommunications "up front."

3. Telecommunications Joins Data Processing As An Equal

A strong trend is the joining of telecommunications and DP/MIS. Former definite lines of demarcation have become nebulous. DP/MIS has become very dependent on telecommunications for distribution of data. Telcom PABX switches now look like computers, with 80% of new PABX development costs being for software.

This integration of telecommunications and DP (computers) is occurring four ways:

A. User staffs are being integrated under Information Management.

B. Vendors are supplying both telecommunications and DP functions, to provide a desirable single source. (E.G., ATT, IBM, etc.)

C. Telecommunications and DP equipment is being installed in the same room.

D. Telecommunications and DP signals are traveling in the same wires.

Granted, DP involves more people than telecommunications, but telecommunications should no longer be considered a "second cousin" to DP as in the past.

4. Need For Network Control Center

All intelligent buildings should be planned to accommodate a telecommunications network control center (NCC). All telecommunications signals must pass through the NCC to be controlled, i.e., switched, measured, maintained, integrated, converted, etc. Failed circuits and equipment must be immediately detected here. Limited staff can be efficiently used. Facilities can be dynamically re-allocated as required. The NCC is the necessary nerve center for controlling increasingly important signals.

The NCC typically houses the on-site digital PABX switch, data communications equipment, multiplexers, modems, test consoles, and other telecommunications equipment. Sometimes the NCC is a separate room and sometimes it is incorporated in the Data Processing Center. Already joining telecommunications and DP in these rooms are systems for environmental control, security, and other building functions. The reason is that these building systems are heavily computer-oriented. Also, operations and maintenance staff can be shared.

5. Sufficient Ducts And Conduits

A big lament of owners and users of older buildings being retrofitted with modern telecommunications is that ducts/conduits are inadequate. That is, they are too small, too congested, and too hard to access. It is not uncommon to pay $1,000 to run a small amount of wiring from Point A to Point B.

The intelligent building must address this important, basic issue, in order to save money, allow rapid installation of systems which increase staff productivity, etc.

6. Need For Solid Planning, Documentation, And Training

These are the three areas which traditionally lose out to the rush of daily operations. They are the first things to be cut when belt tightening occurs. A few quick dollars can be saved for a short while, but the long-run cost to the intelligent building will be much greater.

It appears this lesson may never be learned. It has not been learned over the past 100 years of telecommunications. The problem is that: money lost later cannot be readily identified with cost cuts made earlier in planning, documentation, and training; the money gets charged to someone else's budget; and the person responsible left the company two years ago.

A particular pitfall is that prior to 1984 divestiture, ATT/Bell performed much of the planning, documentation, and training for building telecommunications. As a result, users who are now "on their own" will underestimate the importance of these three functions. For example, two years down the line, a major system (PABX, T-1 link, etc.) will fail and a two-minute repair will take half a day because documentation is poor, no training was provided, and inadequate emergency back-up was planned.

Probably, the managers of intelligent buildings will make some of the same mistakes. The only hope is that the broad, across-the-board integration of functions in intelligent buildings may make the cost of weak planning, for example, very apparent, with definite cause-and-effect immediately obvious.

7. Design Flexibility Into Telecommunications Planning

Intelligence in buildings is meant to increase the profitability and extend the lifetime of buildings. Much will depend on how well the building designers anticipate advances in future technology. This is especially true for telecommunications.

There is enough experience now to realize that designers should not rely on any one telecommunications technology, no matter how promising it may look today. For example, millimeter waveguide was a very promising transmission medium years ago, but it lost out to fiber optics and other media.

The best advice is to design in flexibility. Treat different transmission media as complementary rather than competive, and allow for all, via extra ducting/conduit. Likewise with circuit switching vs packet switching. The returns are not yet in for telecommunications technology. We have only just begun.

The key is to select the right "common denominators." For example, going digital provides much flexibility; and, in particular, going digital T-carrier, 64 Kilobits/sec, and so forth. All transmission media and all telecommunications switching offer these standard units, i.e., common denominators, which will be used well into the future, regardless of technology.

A designer of intelligent buildings might, for example, gamble on fiber optics to be the transmission medium of the future, and save some money by pre-wiring the whole building with fiber exclusively. If future technology proves the designer wrong, the 20% saved initially will prove costly later. The advice, then, is to pay the extra 20% now to achieve the telecommunications flexibility likely needed later.

8. Plan Properly For Tenant Telecommunications

The developers of intelligent buildings will have some big decisions to make regarding tenant telecommunications. Should they provide all telecommunications and require tenants to use it? Should they provide just the potential and later install what tenants request? Should they provide a bare minimum (say ducting/conduit only) and let tenants go it on their own?

These are major, strategic decisions; and they should be addressed before building design commences. This would require surveys of prospective tenants. The developers of the intelligent buildings must recognize there are many types of telecommunications user tenants, such as:

- Tenants who are small vs. large in size.

- Tenants with small vs. large telecommunications needs.

- Tenants with sophisticated vs. non-sophisticated telecommunications needs.

- Tenants who do not seek telecommunications control vs. tenants who want tight telecommunications control.

- Tenants who are forgiving of temporarily poor telecommunications service vs. non-forgiving tenants.

Intelligent building developers must carefully decide the telecommunications profile of the tenants they seek and then carefully survey prospects. Here are some examples of what can go wrong if this is not done carefully.

An intelligent building developer chose a telecommunications strategy whereby all tenants must accept building-provided telecommunications. This developer had confidence in his telecommunications skills and also saw much profit from telecommunications services. He then provided heavily for telecommunications, only to find out the tenants he seeks want the option of providing their own telecommunications, a fact which a survey would have uncovered.

Another intelligent building developer did similarly, but made a survey showing tenants preferred building-supplied telecommunications, although they also wanted the option to provide their own, later. All went well for several years until the building telecommunications service declined in quality due to new, weak telecommunications management. The tenants were non-forgiving and installed their own telecommunications systems, so that they would never return to building-provided telecommunications. As a result, developer's telecommunications investment did not pay off.

The intelligent building developer, then, must recognize that any telecommunications service provided to tenants must be of consistently good quality. This makes even more imperative the proper management of telecommunications, including the planning, documentation, and training earlier described, as well as other good telecommunications management principles. It must always be remembered that intelligent buildings are servants of human needs; that technology must not be applied just for technology's sake.

ADJUSTMENT PERIOD

Intelligent building's people should not be surprised if they encounter difficulties with telecommunications. In fact, they should be realistic and expect to experience some difficulties, as do many users. The key is to plan ahead and minimize problems. It can be done.

Organizations traditionally involved heavily in telecommunications, namely, common carriers, vendors, and users, are all re-adjusting to "the new telecommunications."

The rules have changed. Telecommunications is a whole new ball game. The factors causing this are those outlined earlier under "Major Trends in Telecommunications." They include ATT/Bell divestiture, rapid advances in technology, oncoming information society, integration of voice/data, users being "on their own," and other factors.

Everyone is in an adjustment period. Common carriers are re-organizing and re-re-organizing as they adjust to their new, non-traditional roles. Vendors are stepping out and expanding into new telecommunications areas, with excitement and some trepidation. Users are forming expanded and sometimes new telecommunications departments, and seeking to convert telecommunications from a "necessary evil" to a true "corporate resource" for providing a competitive edge. They are all seeking to grab hold of the new telecommunications with its new set of rules.

The developers of intelligent buildings, who are actually a user/vendor/common carrier combination all rolled into one, are subject to the new telecommunications rules and volatility. Overlaid on this are the unique aspects of applying telecommunications to the intelligent building. There fortunately have been a number of leading-edge developers whose experiences are described in other sections of this book, who are considered first-wave, trial-and-error developers. Additional experimentation will take place over the coming years as intelligent building concepts are refined. Certainly, today's second-wave developers will benefit from others' earlier experiences.

This all is meant not to discourage developers of intelligent buildings, but to encourage them to treat telecommunications with the seriousness and respect it deserves. It is not a casual play toy, to be nickel-and-dimed.

There is excitement and true opportunities for those wishing to step out and be different and play leadership roles, as well as for those preferring to be more conservative, in the management of telecommunications in the intelligent building arena.

Telecommunications Options

There are several ways for intelligent building developers to manage telecommunications. All have their pro's/con's.

1. Hand off the major effort to an established telecommunications organization.

This could be: a common carrier which has established an intelligent buildings group; a vendor which has done likewise, often by acquiring and integrating a telecommunications group with traditional building groups such as enviromental control, people movement, etc.; a system integrator house which manufactures no systems but will provide wide-ranging intelligent building consulting expertise including telecommunications; possibly others.

2. Work with an architect who has added a telecommunications group or will subcontract to a telecommunications consultant.

3. Form a strong in-house telecommunications group.

The first and third options can provide telecommunications management for both the planning and operations phases of telecommunications. The second option, one of planning, will require the developer to arrange for follow-on telecommunications operations via some other means.

As stated earlier under "Dedicated Telecommunications Staff," no matter what option is selected, the intelligent building people should have some in-house staff to establish and maintain control of telecommunications. The wise, confident supplier prefers this, and the less wise and confident needs this.

CONCLUSIONS

Telecommunications must be a major consideration in the intelligent building. Telecommunications is the latest newcomer on the scene. For the building's intelligence to be of much use, it must be distributed. This is done via telecommunications's two basic functions -- transmission and switching.

People involved in intelligent buildings probably know less about the new telecommunications than any other building function. This limitation fortunately can be overcome by following the advice provided in this chapter.

There is money to be made and other advantages to be gained by properly managing telecommunications.

SECTION II

Building Design: Issues And Trends For Intelligent Buildings

Chapter 9

Building Design Issues Resulting From the ORBIT 1 and ORBIT 2[1] Studies

Gerald Davis, AIA
President, TEAG, Inc. - The Environmental Analysis Group

The Challenge

In North America, because of an explosion of information technology and the changing corporate organization, few buildings yield their full potential for owners or occupants. The new technology places physical demands on buildings, but it also brings opportunity for improved facility performance and financial return. Landlords are challenged to use computers to manage and operate buildings more efficiently, to retail telecommunications and other data services to their tenants, and to accommodate and support the hardware that tenants use.

As occupant organizations acquire information technology, they change the demands they place on buildings. When aided by computers, individual managers find they can lead more people, and they reduce the number of levels between top and bottom of the hierarchy. The roles of support staff change, office shift work becomes more common, staff are relocated more frequently, and the facilities get rearranged.

[1]ORBIT-2 was a joint effort by four individuals who were also the principal authors: Gerald Davis as project director, and Franklin Becker, Francis Duffy and Bill Sims. Davis, in preparing this chapter, expresses his appreciation for their collaboration in and essential contributions to the study, and for permission to draw on the study report in the preparation of this chapter.

This chapter summarizes the findings from ORBIT-2 that are relevant to intelligent buildings. To the continuum of building intelligence outlined by Richard Neubauer in his "Overview", this chapter adds a fourth type of building intelligence: <u>capacity for change</u>. For a building to be fully "intelligent" in this sense, it needs to be planned and equipped to accommodate the changed requirements as existing occupants and their technology place new demands on the facilities they occupy, and as potential occupants (or tenants being solicited by a building owner) compare buildings to determine which will best accommodate their information technology and organizational change.

ORBIT-1

Among the first to recognize this, and do something about it, were three firms based in London, England. Francis Duffy, senior partner of the architectural firm, Duffy Eley Giffone Worthington (DEGW), noticed that computers were becoming more and more demanding, while David Firnberg, managing director of the computer consulting firm, Eosys, found that his clients' buildings were often poorly suited to accommodate the computers he was helping them select and install. In 1982, they teamed with Peter Ellis and Sheena Wilson of Building Use Studies to launch a year-long multi-sponsor study of the phenomenon, to learn how to avoid or cope with the problems they were experiencing. The study became known as ORBIT (<u>O</u>ffice <u>R</u>esearch into <u>B</u>uildings and <u>I</u>nformation <u>T</u>echnology.)[2]

That first ORBIT study, completed in March, 1983, explained the impact that new technology was having on design of offices in England, and indicated how this would increase over the next ten years. It developed an expert-based method

[2]Sponsors during the British ORBIT study were: Bovis Construction Ltd.; British Telecom; (British) Department of Industry; Fletcher King; Greycoat City Offices PLC with Norwich Union Insurance Group; Jones Lang Wootton; Matthew Hall Mechanical & Electrical Engineers Ltd.; Scottish New Towns: Cumbernauld, Glenrothes and Livingston; Steelcase Strafor (UK) Ltd.; and Steelcase Inc. (USA). After the study was completed, additional organizations paid the the sponsorship fee to gain access to the findings.

for evaluating existing and proposed buildings, and applied it to office buildings of various ages, sizes, shapes, heights and plan types, occupied by a variety of organizational types and functions. It gave those responsible for the design, provision and use of office buildings in England an understanding of the future trends in information technology, and guidance on how to avoid costly mistakes.

ORBIT-2

The findings from that ORBIT study could not be applied directly in North America. The basic issues are the same, but construction practices, building codes, and office buildings are somewhat different from those in England, as are many organizational practices and worker expectations. Therefore another multi-sponsor study, ORBIT-2[3] was launched by Gerald Davis and Francoise Szigeti of TEAG - The Environmental Analysis Group, Ltd., with the collaboration of Francis Duffy, who had led the British ORBIT study, and Franklin Becker and Bill Sims from Facilities Research Associates.[4] In North America, the word ORBIT referred to ORganizations, Buildings and Information Technology, because of the importance of organizational change under the impact of information technology.

[3]Sponsors for ORBIT-2 were: Alberta Public Works, Supply and Services; Arthur Young; Birtcher; Building Owners and Managers Institute International; Donn Corporation; Exxon Corporation; Honeywell, Inc.; INNOVA; Interface Flooring Systems, Inc.; Mead; Mobil Corporation; Public Works Canada; Steelcase, Inc., Sunarhauserman; Tate Architectural Products, Inc.; TRW Space and Electronics, Inc.; The World Bank; and Xerox Corporation.

[4]Harbinger Group Inc. (then a newly formed, wholly-owned subsidiary of Xerox Corporation) replaced TEAG as prime consultant when Davis, TEAG's president, became a founding principal at Harbinger in 1984. The ORBIT-2 report was copyrighted by the three firms acting together: Harbinger Group Inc., Duffy Eley Giffone Worthington, of London, England, and Facilities Research Associates, Inc., of Ithaca, NY. In 1986, after the ORBIT-2 study was completed, Davis returned to TEAG. Later that year, Xerox closed the Harbinger Group and sold the name to another Harbinger principal, Carlos Santiago, who manages it as a telecommunications consultancy.

ORBIT-2 analyzed trends and the implications of change in organizations, human factors, information technology, the office environment and buildings, facility management and real estate, and the corporate response to these trends. It developed a process for analyzing the present demands of an organization, to predict the pattern of change it is likely to experience in the future, and the capacity of its facilities to meet those shifting demands. It suggested strategies for how buildings should respond to changing organizations as they cope with information technology.

And it recommended facility management strategies best suited for various types of organizations.

Focus On Intelligent Buildings

This chapter summarizes the findings from ORBIT-2 that are relevant to intelligent buildings. ORBIT-2 considered the demands on buildings because of new telecommunications systems, office automation systems and other information technologies; and considered building automation systems as an important aspect of building service systems. ORBIT-2 also considered how the existence of intelligent buildings will changed the way organizations, large and small, conduct their business.

ISSUES BECAUSE OF ORGANIZATIONAL CHANGE

Trends

At the level of the whole organization, ORBIT-2 predicted new organizational structures and procedures, more rapid organizational change, more information specialists, and said that as staff become more dispersed, information technology will be used for more links and monitoring of work and workers. To respond, office facilities should be: more easily adapted and replanned for different occupants; more easily subdivided and expanded; better able to handle electronic systems; better able to accommodate diverse hierarchical and social structures; better equipped with meeting and social spaces; supportive of unplanned, informal, face-to-face contacts; and well equipped with technical support services.

Within organizations, small and mid-sized groups will become increasingly significant, as opposed to the larger branch or department. Facilities will be planned and laid out to reinforce group identity, for example with individual workstations or offices arranged around shared or group spaces rather than being arrayed linearly along an unzoned corridor or aisle. This will create demand for furniture and fittings appropriate for various group sizes and work styles, and electrical, electronic and HVAC servicing systems that conveniently support group work. This will include electronic file servers and local area networks that link individuals into small groups, and then link these small groups to other groups. Space planning will provide more and better spaces for work by project teams and other group work, within the space assigned to small groups.

Each individual's need for more control of air, temperature, furniture, lighting and other physical aspects of the work-station will be recognized in building design and furnishing. Multiple work settings will be provided for staff who work on several different subjects, requiring different reference papers or electronic instruments, during the course of a typical day.

The Key Issues From ORBIT-2 About Organizational Change

ORBIT-2 identified nine key questions about a specific organization that managers should ask when determining the relative importance, for that organization, of various features of an office facility.

Change of total staff size: Is the total number of people to be accommodated in the organization stable, or changing?

Attract or retain work force: How important is it for the organization's success to ensure that highly qualified staff who are hard to replace feel satisfied enough not only to come but also to stay?

Communication of hierarchy, status and power: How important is it for people to recognize differences in rank, status, power within the organization?

Relocation of staff: How frequently are people physically relocated from one workplace location to another inside the office?

Maximizing informal interaction: How important are informal and spontaneous interaction and face-to-face communication among staff?

Human factors for the ambient environment: How important to the organization is the quality of lighting, air conditioning, air quality, temperature, acoustics, etc.?

Image to the outside: How important is the image of the organization which is presented to visitors from the outside?

Security to the outside: How important is protection of information and other valuable objects from outsiders?

Security to the inside: How important is the protection of information from insiders?

ISSUES BECAUSE OF INFORMATION TECHNOLOGY

Myths, Realities And Consequences

When looking at the future of information technology and its effect on organizations, the ORBIT-2 study identified several myths that were leading managers to unwise decisions about their office facilities.

The permanence of change. Myth: Once most people in an organization have a desktop personal computer or terminal, information technology will no longer be an issue that managers have to deal with. Reality: Change accelerates. Equipment is replaced. Networks are adapted. Offices are refitted.

Information technology affects everyone. Myth: Information technology is for professionals and junior staff, not managers. Reality: Tomorrow's managers will use information technology themselves, but most of today's older managers will avoid using a keyboard.

Environmental impact. Myth: The problem of heat generated by computers and word processors has been eliminated by new low-power equipment. Reality: The proliferation of computer devices will offset the decline in power used by each, at least in the near future; also, the trend will be back to higher illumination at the task. Clusters of file servers and printers cause hot spots.

The end of meetings. Myth: Electronic mail and video conferencing will make meetings obsolete. Reality: The great increase in available information increases the hunger for face-to-face contact with peers, and more meetings occur.

Saving time. Myth: Information technology saves time. Reality: Some tasks take less time, but more time is often needed to find the right format, or to finish an expanded task made practical by the new technology. The office day is lengthened to exploit the costly equipment, but people no longer have to be on the spot all the time. Thus, time is used differently rather than saved.

Technology solves its own problems. Myth: Problems solve themselves because equipment steadily needs less power and less space, while giving off less heat. Reality: With more powerful equipment, the heat and noise often either increase or remain about the same, and more machines are introduced. Older equipment is rarely discarded, but is handed down or used as back-up.

The compact, paperless office. Myth: Computers save space and eliminate paper. Reality: Computers can reduce paper-based memoranda that keep an organization's processes running smoothly, but printers and photocopiers will be busier than ever. People will still need workspace for correspondence documents and for business discussions, and browse through books, articles and files.

Space for electronic equipment, printers and print-outs is extra.

Equipment can work anywhere. Myth: The new small computer technology can be fitted into any office. Reality: All equipment makes demands on the building: acoustics, lighting, security, protection from static and dust, power, humidity, temperature, fire protection, connection to cables, protection from power surges.

Cable management. Myth: Local Area Networks (LAN's) will solve the cabling problem. Reality: More cables and more connections will be needed by many organizations for at least five to ten years. This offers a challenge to manufacturers of furnishings and partitions and is a boon to makers of access (raised) floors, and other innovative solutions.

The Key Issue From ORBIT-2 About Information Technology

ORBIT-2 identified nine key questions about the information technology used by a specific organization that managers should ask when determining the relative importance, for that organization, of various features of an office facility.

Connecting equipment: How important is it that all or most electronic workstations be connected to networks, main frames, or other electronic equipment not located at the workstation?

Changing location of cables: How important is it to be able to easily change where cables end, that connect electronic equipment, whether at workstation or in equipment room?

Environmentally demanding equipment: How important is it to meet special environmental requirements, such as for cooling, floor load capacity, humidity, dust, acoustics, vibration, static control, etc., due to the presence of information technology equipment?

Protecting hardware operations: How important is it for operations not to be interrupted, even for a few seconds; or for data to be protected against loss, delay, change or misrecording, due to problems with computer or related hardware, e.g., due to interruption of electrical power, or other hazard?

Demand for power: How important is the need for primary and secondary electrical power capacity and feed, including vertical and horizontal on-floor distribution?

Relocating heat producing equipment: How important is it to be able to relocate heat producing information technology equipment to unpredictable locations within the office work areas?

Human factors at the workstation: How important is the provision of suitable workstations, with ergonomically appropriate furniture, equipment and task illumination, and sufficient horizontal and vertical space or all required information technology equipment?

Telecommunications to or from the outside: How dependent is the organization on large volumes of uninterrupted telecommunications to or from outside the building (including voice data or video, etc.)?

RECOMMENDATIONS FOR THE BUILDING SHELL

Location And Tenure

Some office users will change their locational requirements, and relocate as their organizations change in consequence of their acquisition of new information technology. For instance, expect shifts of tenant demand from downtown to suburban sites because of an increasing user preference for avoiding the city center with its higher costs for housing, commuting and office facilities. At the same time, those organizations which give high priority to "attract or retain workforce" typically need to be located near off-site amenities, such as options for shopping and convenient computer transportation.

Expect changes also in the ratio of owned to leased property for large corporations, reflecting the increasing frequency and extent of organizational change, a growing recognition of the financial potential of real property assets, which should perform or be sold, and changes in the taxation and economics of real and movable property.

Building Form And Structure

Plan for: bigger floor to ceiling heights; fewer columns and longer spans; floor plans in which more workers have a view to the outside, (less deep space where workers are far from a window); increasing emphasis on horizontal rather than vertical buildings, to allow large total continuous/connected floor areas; building placement and configuration planned to conserve energy; and demountable building skins which are capable of being changed several times during a building's life.

RECOMMENDATIONS FOR BUILDING SERVICES

Electrical And Electronic Systems

Plan for increased watts per square foot to serve information technology (organizations with one personal computer for each worker plus a file server and multiple printers for each work group are probably close to their peak electrical power demand, but in 1987 most organizations have only a fraction of that.

In new and rehabilitated buildings, the trend is either to provide more intelligence (both building and user), or to provide more places for equipment and the paths for connecting them. Cable management becomes critical, particularly when there is a change-over from one network or mainframe to another, and the system must be up, running and tested, and staff trained on it in parallel with the old system, before the old one can be shut down and its cables removed.

To facilitate cable management and reduce on-floor runs, cable risers and distribution closets will increasingly be dispersed. In buildings with only one core and large floors, they will often migrate to locations near the perimeter.

In the 1990's expect increased emphasis on data security in buildings, which makes fiber optic cable more attractive and will lead to more shielding of electromagnetic emissions, even to "Faraday"-shielded rooms. (Do not expect optical computers for non-military applications before the late 1990's, and likely later.)

Integrated design of structure and building service systems at an early stage of building design will lead to more permeable building forms, better suited to support the frequent changes and system integration of fully intelligent buildings. Linking of information system management with building system management will be more typical in intelligent buildings owned by the occupant organization, and where security of information and building can be under a single control.

Heating, Ventilating and Air Conditioning Systems (HVAC)

As buildings become more intelligent and control systems more integrated, HVAC systems will be more responsive to group and individual requirements, and provide automatic response to changing demands over the 24-hour day in small areas of a floor, or individual workspaces. Prime movers, such as boilers, chillers and cooling towers, will still tend to be centralized, but air handling units will increasingly be dispersed for economical and easy response to frequent changes in the zoning, levels and timing of demand for heating, cooling or fresh air on the office floors.

To reduce incubation and recycling of bacteria, viruses and fungi by air handling systems, today's water-spray humidification systems, important both for static control and employee health, will be replaced with other technologies for keeping moisture in the office air, and filtration systems will be improved.

Illumination

In new buildings, despite reductions in solar gain, the exterior envelope will admit outdoor light deep into the interior.

As visual display technology evolves during the mid-1990's, illumination levels both on the task and for the overall workstation will increase, but illumination of the ambient office environment will continue to be varied and at lower levels than were recommended in the 1960's. Illumination sources will continue to migrate from architecture to furniture.

Acoustics

Fan noise will be the bane of the 1990's office, as typewriter noise was in the 1970's. Freedom from distracting sound will be increasingly important for knowledge workers, and together with the need for speech privacy will lead to greater use of enclosed offices (and additional options for cable management.)

RECOMMENDATIONS FOR FITTING OUT

Accommodating Change

The traditional boundaries between furnishings and fixtures are dissolving as illumination becomes a furniture item, air supply moves through the floor into furniture for discharge, furniture shelves hang on gypsum drywall partitions from the same brackets as on furniture systems, and furniture screen systems grow into full-height walls. Intelligent buildings housing knowledge workers with dense information technology will be subject to increasingly frequent internal rearrangement, and shorter life-cycles for an increasing proportion of the total investment in the office environment.

Most new information technology will be installed in existing buildings; therefore much "intelligent building" investment will go into retrofitting. This is often more difficult in recently-built offices than in older buildings with robust structures and large floor-to-ceiling heights that can accommodate cables, air ducts and access floors. Even harder is keeping buildings operating and occupants productively at work during rehabilitation, despite the dirt, noise and interruption of building services that typically accompany construction.

Ancillary Facilities

Offices have long contained conference rooms, machine rooms, stock rooms, coffee break areas and other support spaces. These facilities will be increasingly important, and will occupy a larger proportion of total floor area. Increasingly, support spaces for work groups will be distinguished from support spaces that serve the entire organization, for example: team work and meeting areas separate from organizational conference centers and training/class rooms; group pause areas separate from organizational food service, etc.

The spaces to which visitors are admitted and in which public image is important will increasingly be separated from the back-stage work areas of small groups and task teams. More and more of the latter, together with their ancillary spaces, will move out of high-rent downtown locations to corporate back offices.

Behavioral Considerations

As office hours change and become more flexible, ancillary spaces will be recognized as particularly important for staff productivity and morale.

Facilities should be designed to help staff identify with their small work groups or project teams; furniture screens or full-height partitions are typical separators. Links to local area networks and clusters of people who cover each other's telephones should coincide with the physical boundaries of their assigned office space. Because the size and membership of these groups will change frequently, demands on a building's "intelligence" will increase. (Most organizations will see increases in their churn rate, and many will relocate half their staff each year.)

Wherever attracting or retaining staff is important, valuable assets will include: individual control and choice over settings at the workplace for HVAC, illumination, and even aspects of acoustics, as well as over settings for each individual's electronic environment. Organizations with this need will have an added reason to find intelligent buildings attractive.

The "status" issue in offices will never go away, and workstation size is a traditional way of expressing status. However, increased standardization of the size of enclosed offices can be expected, with workstations for lower-level workers gaining space, adding work surface and enhancing enclosure. At the same time, there may be some trading off of individual small workstations for space in shared group areas for common activities.

Where people do various kinds of tasks during the course of a day, some organizations with compact professional teams will create multiple workplaces, one for each kind of task, clustered within a shared group area.

Ever-changing organizations need to seem stable to their members and to visitors. It is essential that people find their way around despite frequent relocations of individuals and work-groups. Building form and office layout, and signage, cues and signals: all can help. One aspect of the intelligent building's potential that has hardly been explored is the possibility for using electronic signs and signals to help define place and routes, and to give added, temporary cues to visitors seeking a particular individual at an unfamiliar location.

RECOMMENDATIONS FOR FURNISHINGS

Systems

Cable management in the zone between the floor, wall or ceiling and the connection point on desktop equipment will continue to be a challenge, although innovative solutions are now emerging from some furniture manufacturers.

Rearranging the office by rearranging the furnishing system has become a complicated and expensive process, requiring skilled labor and for some systems an array of special tools and spare parts. To solve this, the trend will be to free-standing furniture units, and to systems in which the office occupant can make rearrangements within his/her workstation without needing special tools or calling for movers.

The demand for filing and storage equipment will not abate, but electronics are changing the kind of information printed on the hard copy that is stored, and therefore where it need be located. As information becomes more readily accessible when archived electronically than by retrieving hard copy from "dead" files, the demand for on-floor and basement group storage will diminish. At the same time, as computers and printers speed production of more drafts and more frequent reports, there will be more, not less paper at the individual workplaces of support, technical and professional staff, up to the manager level.

Human Factors At The Workplace

Work with information systems often introduces new technological stressors, such as glare, veiling reflections, flicker and jitter with visual display units, and electronic pacing and monitoring of performance, that increase the overall stress on office workers. If these cannot be avoided, then to keep the overall level of stress from reducing worker productivity and impairing health, other sources of potential stress should be reduced. In some workplaces there is potential for achieving this balance by removing physical or behavioral stressors from the physical environment.

For example, in many organizations there is a new emphasis on allowing individual staff members to choose what goes into their workplace. Stress is reduced when people have greater opportunity to personalize their individual and group territory, and adapt it not only to their task needs but also to their personal taste. Reducing intrusive or distracting noise, improving ambient and task illumination, realigning workstations to reduce glare and excessive contrast in the field of vision are among the many factors to consider.

Deciding what should be changed, and what stressors should be tolerated by workers, requires understanding the wide range of individual differences among workers. If one or two workers in a group complain vociferously about aspects of the physical work environment or the performance of computer technology, while others experience no problems, managers should not assume that the complainers are malingering.

The variation of human response to office stressors is much wider than generally understood, and often much greater than typical differences due to age, sex or skill level. For instance, in the author's personal experience, a 60-year old male had a "fast eye", and was seriously impaired by flicker of a video display that was easily tolerated by most 20-year olds; and a secretary had her productivity reduced by working under normal fluorescent illumination acceptable to her co-workers, and was hospitalized after working for only an hour at a typical light-table illuminated by normal fluorescent tubes, because of the flicker.

Differences in worker size, weight, length of arm reach, height of eye, response to heat, cool and air movement, acuity of vision, alertness, and other physical characteristics: all may need to be accommodated when workers do repetitive tasks, such as is common at computer terminals. To achieve full productivity when equipment is used for multiple shifts, the seating and other aspects of the physical setting should be easily adjustable when a different worker takes over at each shift change.

RECOMMENDATIONS FOR FACILITY MANAGEMENT

Facility Management Strategies

Facility management is becoming recognized as important for achieving full organizational effectiveness. Major corporations such as Aetna Insurance, and institutions such as the World Bank, have brought facility management together with the management of office and information systems, reporting to the same senior executive. This general trend is further detailed in Chapter 29. Concurrently, facility management is becoming professionalized, with its own professional association. Several universities and colleges offer graduate and undergraduate programs in the field.[5] The goal of facility management, according to the International Facility Management Association (IFMA) is "providing and maintaining human and effective work places, where both productivity and human factors are considered."

Hence, to cope with the changes that information technology is bringing to the office, facility managers will need more than technical skills. They will also need a very good understanding of how the physical environment affects behavior, motivation, performance and job satisfaction.

Facility managers will become involved in corporate planning, as they manage assets that for the Fortune 500 companies total over a quarter of all corporate assets. They need

[5]At this writing, degree programs are in place at the Department of Design and Environmental Analysis, Cornell University (the pioneer), the Program in Social Ecology at the University of California at Irvine, the Department of Human Environment and Design at Michigan State University, the School of Business at Grand Valley State College, and others have been announced.

skills, knowledge and concepts from architecture and interior design, engineering, human factors and environmental psychology, business and real estate, building construction and management. Benchmarking of facility operations is likely to increase. Firms will want to compare aspects of facility performance such as costs of occupancy, costs of change, density of space utilization and employee satisfaction.

CONCLUSION

Choosing Which Facility To Occupy Or To Acquire Or Hold

Decisions on facilities no longer focus on such relatively simple factors as image or acquisition costs. Will the organization be making the information revolution work for it, or will it be fighting a rear-guard action against its competition and its staff? Will huge investments in hardware require changes in how and where a company conducts its business? Will a new kind of labor force work in harmonious productivity, or will management be facing unprecedented staff problems which are hard to fathom? Will far too much, or far to little, be tied up in concrete, in workstations, in equipment.

Rating Of Facilities

Buildings that look the same rarely perform the same. The breakthrough in the ORBIT-2 study was the discovery that objective guides and comparative rating of facilities are possible. Buildings can be mathematically rated much more simply and surely than people can be. Facility management also can be rated, and matched for effectiveness against the organization it supports.

Decision-makers can use the rating process to compare options and make choices about intelligent buildings (as distinct from the pass-fail type of evaluation needed for life-safety and building code approvals.) The rating process is now being standardized through ASTM Subcommittee E06.25 on Overall Performance of Buildings, where a task group is translating it into standard ASTM format and putting it through the consensus approval process. In its present form it is already available to participants in the Subcommittee, which is open to all and welcomes volunteers.

Looking At The Future

As the rate of change in organizations accelerates, staff and decision-makers need to be informed, and awareness heightened at all levels, of the nature of the problems - and opportunities - which they face in realizing the highest potential of their working environment.

As never before, even apparently trivial decisions about the fitting out of an office need to be guided by the best possible estimates of future directions. Buildings are relatively permanent. Organizations of people are not, and the technology that people invent is likely to be the least static of all. The contradictions which inevitably exist between structures and what or who is in them should be the perspective in which current decisions are made. Otherwise, those decisions may be transitory, ad hoc and potentially costly.

Chapter 10

HVAC, Lighting And Other Design Trends In Intelligent Buildings

Fred S. Dubin, P.E.
Dubin-Bloome Associates

The intelligent building is normally defined in terms of state-of-the-market building automation, information processing and telecommunications services.

However, most of the buildings which are defined as intelligent are not really so smart because they are deficient in two broad areas. One of these areas is the building envelope and plan which are not as smart as the electronic systems that they enclose.

The other area is mechanical, electrical and structural systems which are not smart because they do not provide the environmental quality required by an intelligent building. Conversely, the mechanical and electrical systems do not adequately serve the special needs of information technology and telecommunications equipment and systems in an intelligent building.

Existing technology and design knowledge are available to address these deficiencies and can be employed by a smart, interdisciplinary, building design team in a cost-effective manner. This interdisciplinary or, better yet, trans-disciplinary team must consist of an architect, mechanical engineer, electrical engineer, structural engineer, acoustical engineer, interior designer, strategic planner, communications consultant, physiologist, epidemiologist, and members of other related disciplines.

The members of this building design team should be people whose professional abilities cross over normal disciplinary lines. They should be people who understand and feel comfortable with each other's expertise.

Overall, the truly intelligent building which this team will design requires:

- A building envelope which reduces heat loss and gain through insulation and thermal mass;

- Moisture control which is achieved by properly locating vapor barriers and external shading devices to reduce glare and solar heat gain;

- Properly-located windows equipped with insulating glass, heat rejection glazing, and light shelves or other devices capable of simultaneously reducing glare and reflecting daylighting deeper into the building's space. This interior use of ambient light is needed to minimize contrast between the building's immediate inside perimeter and occupied spaces 20 to 30 feet from the perimeter.

- Double or triple panes of glazing which are necessary to prevent condensation on windows due to the higher humidity levels required for proper operation of electronic equipment.

- Vertical shafts for duct work and power and communication cables which are larger than normal from the standpoint of past practices and strategically located to reduce horizontal distribution runs.

- Telephone and electric closets on each floor which are adequately sized and ventilated or cooled to protect equipment.

Building design must meet these requirements in an intelligent building because it is the first line of defense in terms of reducing energy requirements. In addition, building configuration, materials and construction influence the type, size and performance of mechanical and electrical systems.

In effect, the building is the catalyst in achieving desirable inter-relationships between thermal, visual and acoustical environments and systems. At the same time, building structure and materials affect worker productivity, aesthetics, maintainability and life cycle costs.

In addition to these building design requirements, there are a number of additional HVAC, lighting and other needs which a structure can and must be designed to cost-effectively satisfy if it is to truly qualify as being an intelligent building.

AIR QUALITY MONITORING AND CONTROL

One of the major requirements for an intelligent building which is not presently being met is an air quality monitoring and control capability. Air quality in office and other types of buildings today is oftentimes very poor, resulting in health problems, reduced productivity and increased absenteeism.

There are many different kinds of pollutants in indoor air, including formaldehyde which is outgassed from building materials, furnishings and glue in carpets; chlorine, sulphur and ammonia products; carbon dioxide and monoxide; contaminants resulting from copying and reproduction operations; and cigarette smoke.

These pollution problems have always been present; but they are currently growing in severity due to the increasing use of new, synthetic materials in buildings and the expanding implementation of energy conservation measures. These measures make buildings tighter to reduce infiltration while minimizing the amount of outside air used in ventilation.

In addition, the problem is exacerbated by the development of the intelligent building since the insulation on cabling used to connect computers and other electronic equipment can be a significant source of outgassing. The level of one toxic material may be low enough to cause no health problem. But the same contaminant in combination with one or more other toxic materials can produce a mixture of compound injurious to health.

Ways To Improve Air Quality

The best way to improve air quality is to minimize if not eliminate pollution at the source wherever possible by specifying building materials, furnishings and equipment which do not outgas contaminants. Another effective approach is to group together copying and printing processes and rooms where cleaning materials are stored so that contaminants can be efficiently exhausted to the outside. Smoking areas with special exhaust capabilities can also be established as necessary. Non-smoking areas are also very desirable and smoking is being banned in more and more buildings.

Outside air intakes should be located away from obvious pollution sources such as exhaust discharge ducts and truck loading docks. Contaminants coming into the building with outside air should be monitored and controlled through filtration, absorption, sprays or electrostatic precipitation, depending on the type of contaminant. In addition, indoor sensors should be used to constantly monitor air quality in terms of the chemical composition and parts per million of each contaminant. However, if indoor air pollution still exceeds predetermined limits, the next step is to control air quality through increased ventilation and improved air distribution patterns.

Use Ventilation To Control Air Quality

The most cost-effective way of providing building air circulation for heating, cooling and ventilation today is through the use of a variable air volume (VAV) system. A VAV system conserves energy by reducing the volume of air delivered to a space based on its thermal requirements, thus reducing power required for fans. A number of different types of VAV systems can be used.

One is a VAV dump system which closes dampers and recirculates air from supply back to return through a ceiling plenum without delivering it to a space when temperature requirements of the space are being met. Since less air is circulated through the duct system, the system's static pressure is reduced and less supply fan horsepower is required.

Another system uses damper supply ducts to build up pressure as a way of reducing air volume. Operating a little more efficiently, a vaned inlet system reduces air intake to the supply fans to achieve the same objective. The return air fan is keyed to the supply fans so that it also reduces the amount of exhaust air, thus saving power. Another VAV system approach involves using a three-speed fan motor to vary air volume. The most efficient system consists of a variable frequency drive which can provide an infinite number of fan speed settings based on space temperature requirements.

All of these VAV systems conserve energy by reducing the amount of ventilation or outside air delivered to a space when permitted by temperature conditions. However, this reduction in air volume has undesirable effects on air distribution patterns.

For example, air may short-circuit from supply diffusers to return air grills. In addition, air does not circulate properly through personnel workstations, particularly if they are built with walls to the floor and fitted with acoustically-absorbent materials or baffles necessary to reduce noise but that further obstruct air movement. As a result, work station air quality suffers because contaminants are neither diluted with outside air nor flushed by overall air flow.

One method of solving this problem is to install power or induction units at outlets of VAV mixing boxes to induce additional room air into the primary air stream. This makes it possible to maintain proper air distribution volumes and patterns around workstations, even though a VAV system may be operating with reduced air volume to conserve energy.

Another approach is to use mixing dampers to deliver an increasing percentage of outside air when air quality sensors indicate contamination buildup while recapturing sensible and latent heat from the exhaust air stream through the use of a heat exchanger. Applicable to cooling as well as heating operations, this method retains the energy-conserving benefits of VAV operations while providing contamination control.

BAS Provides Ultimate Control

Air quality can be controlled using multiple, small air handling units, each of which varies the amount of fresh air delivered to a space based on input from pollution sensors installed at floor, zone, office and even individual workstation levels.

But what if indoor pollution rises to an unacceptable level, even though outside air dampers are opened to provide 100-percent outside air? In this case, VAV system fan speed would have to be increased to bring in even more outside air. At the same time, it would be necessary to have mixing boxes equipped with terminal heating and cooling coils to respectively prevent overheating and overcooling. Furthermore, if outdoor air quality were subnormal, additional air treatment would have to be provided rather than bringing in more outdoor air.

Ultimately, the intelligent building must therefore have a Building Automation System (BAS) capable of sensing and optimizing all of the environmental conditions or factors involved in achieving required air quality. These factors include not only temperature, humidity and energy consumption but new inputs such as air quality, air movement and number of building occupants. Equipment and controls required to implement this building automation system are available today.

The building automation system will optimize all of these factors to achieve the best possible results while insuring that required air quality is maintained. This will result in a better work environment while improving personnel health, safety and productivity. In addition, it will provide an improved operating environment for the proliferating use of electronic equipment in the intelligent building.

DESIGN FOR INCREASING LOADS

The intelligent building's use of electronic equipment such as workstations is expanding rapidly. Some intelligent buildings currently provide workstations in the form of video display terminals, personal computers and other input/output devices for as many as two out of every three personnel. Eventually, this workstation/personnel ratio will become one-to-one.

As a result, intelligent buildings must be designed to handle much greater electrical loads. Five years ago, for example, conventional buildings were designed to provide 1-1/2 watts/sq. ft. Today, similar buildings are being designed with 2 to 3 watts/sq. ft. But an intelligent building requires 6, 8 or even 10 watts/sq. ft. in addition to lighting needs and 20 to 30 watts/sq. ft. at workstation locations.

Workstation growth is also resulting in fast-growing heat loads because each workstation is equivalent to a heat load of four to eight people. Conventional buildings are normally designed with one ton of refrigeration for every 600 sq. ft. But, based on a one-to-one workstation per person ratio, the intelligent building needs one ton of refrigeration for every 200 sq. ft. an increase of 1.2 CFM (cubic feet per minute) per sq. ft.

In addition, it is anticipated that the "energy crisis" will re-emerge in future years because the current oil glut and stable prices are only temporary. Prices will rise as world economies recover, fossil fuels will once again be in short supply and the future of nuclear power will remain uncertain.

Consequently, due to growing heat loads and rising energy prices, it will become increasingly more important to use a building automation system to increase the efficiency of the building envelope and HVAC system operations in intelligent buildings. At the same time, other improvements will be made through the use of more efficient alternative energy and power sources.

Increased HVAC System Efficiency

Increased HVAC system efficiency can be achieved through the use of more sophisticated building automation system control programs. Currently, for example, cooling operations are typically controlled on the basis of chilled water temperature and heat load. The control objective is to use warmer chilled water for cooling as loads decrease because this saves energy.

However, when chilled water is warmer, more air must be used to achieve the same cooling level. This results in increased energy consumption in terms of fan horsepower and duct losses which normally are not considered in HVAC system control operations.

As a result, there is a need for a more sophisticated control program which will further lower energy consumption by optimizing chilled water temperature, fan horsepower and duct losses. Other variables which should also be optimized by this program include chilled water pump horsepower and piping losses. In addition, the program should balance these variables with air quality and ventilation requirements. Expert systems are now being developed which will be capable of preparing this program.

The use of cooling storage is another way in which significant improvements can be made in HVAC system efficiency. Cooling storage is accomplished by economically operating a chilled water or ice-making plant at night during off-peak hours and with lower condenser temperatures. The following day, the stored chilled water or ice is circulated through cooling coils in air handling systems or fan-coil units without running chillers, resulting in reductions in operating costs and peak electricity demand.

The HVAC system for an intelligent building must provide greater humidification to reduce the possibility of static electricity which can cause electronic equipment to malfunction or even fail. At the same time, cooling efficiency can be increased by as much as 20 percent by the use of dessicant when dehumidification is required. Absorbing rather than condensing moisture, dessicant use reduces the amount of chilled water required and permits the use of higher chilled water temperatures since it eliminates the need to achieve dehumidification through cooling.

The multiple use of equipment is an additional way of achieving increased HVAC system efficiency. An example of this is the use of a heat exchanger as an indirect evaporative cooling system.

Alternative Energy Sources

The use of solar energy for heating new intelligent buildings will increase rapidly in the future. Conserving energy while reducing heating costs, solar energy uses major passive techniques such as direct gain, trombe walls and sun spaces including greenhouses and atria. Building envelopes will also be designed to do a better job of controlling heat loss or gain. In this regard, building materials can provide both shelter and energy storage, each of which can be controlled as necessary.

Photovoltaic cells are being used for direct conversion of solar energy into electricity primarily in remote and specialized installations. However, conversion efficiency will continue to increase while costs come down from the current $6 to $2 per peak watt. This will make photovoltaic electricity generation competitive with other electrical energy sources, triggering a huge increase in installations.

Intelligent buildings should therefore be designed with a built-in capability for installing photovoltaic cells in terms of orientation and slope of outside surfaces. In addition, electrical systems must be designed for compatibility with alternative energy sources such as photovoltaics, and HVAC systems must be designed for compatibility with solar thermal collectors. Low-temperature media in the heating system permit the use of low-cost solar collectors which provide reduced hot water temperatures or air for heating.

Power Sources

Co-generation or on-site electric power generation will be a growing factor in intelligent buildings because it makes it possible to more efficiently handle increasing electrical and heat loads and generate different-voltage power as needed. It also has the capability to produce cleaner power without spikes, resulting in smoother operations and fewer malfunctions in electronic equipment. Furthermore, more efficient absorption cooling units that use lower temperature fluids for the generator will permit the use of less costly solar collectors.

It is also important to correct the power factor if it is below .95 since this would indicate that the system is paying for power that is not being utilized. This adversely affects performance and operating costs of motors and equipment.

Technology is now available to automatically correct the power factor when it changes in response to load and the addition of equipment and motors with inherently poor power factor. In addition, an uninterrupted power source is necessary so that computer operation and storage integrity are maintained. Batteries are not only the most reliable source of uninterrupted power but eliminate the delays which are experienced with diesel generators used for emergency power. Adequate ventilation must also be provided as required.

DESIGN FOR MOVING LOADS

Electrical and heating loads in intelligent buildings are not only increasing; they are also constantly moving. This is because organizational changes result in continual personnel relocations. In some cases, changes may affect as many as half of all employees per year. As these people move, their workstations and related loads go with them.

Intelligent buildings therefore must be designed with the adaptability necessary to accommodate these moving loads. Based on this requirement, there is a trend to handling base loads centrally while using decentralized, unitary equipment to meet needs created by moving loads. Installed as an integral part of workstations, this equipment consists of miniaturized, local HVAC units with cooling, heating, humidification, exhaust, ventilation and control capabilities.

These HVAC units can be served by duct work and power and communications cabling which is installed either above a hung ceiling or below a raised floor. In the past, the hung ceiling has primarily been used for duct and cabling installation. But, because of increasing load movements, there is a trend today toward use of the raised floor.

Use of Raised Floor

The raised floor has been used for many years to accommodate cables and cool air distribution systems in computer rooms. More recently, raised floors have been installed in large office areas. The floors are initially used to carry cables and then to transport warm or cold air supply and return distribution.

Raised floors provide an adaptability to moving loads which is far beyond that of hung ceilings. The floors can be designed with a "tap-in" capability so that workstation duct and cable connections can be simply plugged in at desired floor locations, eliminating the need for ceiling drops or power poles.

The complete availability of all services at floor level directly under workstations provides a number of benefits. HVAC system efficiency and effectiveness are improved, for example, because air does not have to be as cool when it comes out of the floor and air movement is located exactly where it is most needed. In addition, installing a raised floor in place of a hung ceiling can be more cost-effective overall.

The construction cost of a raised floor is $6 to $8/sq. ft. in comparison to $2/sq. ft. for a hung ceiling. Use of a raised floor therefore results in a net addition to costs of $6/sq. ft.

However, a raised floor can be built with less depth than a hung ceiling, making it possible to reduce floor-to-floor building heights. Moreover, the installation of an HVAC system in a raised floor is easier and less costly.

These savings can pay for the added construction cost of a raised floor, resulting in the same or even lower first costs. Most importantly, costs involved in making changes to accommodate moving loads can be substantially reduced since a raised floor permits changes to be accomplished faster, easier and at less cost.

GREATER INDIVIDUAL CONTROL

Greater individual control of HVAC system operations is another trend which will become increasingly more evident in intelligent buildings equipped with unitary workstations. This will enable building occupants to completely control heating, cooling, air movement, filtration and even humidification, if desired, at the workstation level. Individual control will be further enhanced through the use of personnel sensors and other control devices capable of automatically activating HVAC system operations.

Individual building occupants could also possibly program their own environment through the use of direct digital control capabilities which are increasingly being implemented in intelligent buildings. Providing simpler operations and faster response times, direct digital controllers are able to sense changes in environmental conditions and immediately and directly operate HVAC devices such as dampers or valves, eliminating the need to send signals back to a central computer to achieve device activation. Manufacturers are now producing packaged air handling units with integral direct digital control capabilities.

MORE EFFICIENT, EFFECTIVE LIGHTING

Intelligent buildings will make growing use of natural daylight illumination to improve lighting quality while reducing energy costs.   Other trends will include increasing implementation of task/ambient lighting and higher-efficiency lamps and other devices.

Use of Daylighting

Daylighting can provide a higher-quality, more attractive ambiance while significantly cutting energy usage, particularly in areas which are primarily used from 8 a.m. to 6 p.m. An entire top floor of a building, for example, can be top-lighted or skylighted, eliminating the need for any other ambient lighting. Light shafts extending from the roof down to lower levels can be used to bring light in and distribute it within a building.

Another daylighting technique is to install light shelves outside windows. Reducing glare at the perimeter, these shelves reflect even, ambient illumination up to 25 or 30 feet into interior areas. Atriums will also be increasingly used to bring in multi-directional, natural lighting as well as solar energy and natural ventilation to interior spaces.

Photocells will play a growing role in reducing electric lighting to predetermined levels when daylight is available. Personnel sensors to control lighting and heating will be increasingly used for control and energy savings. In addition, increasing use will undoubtedly be made of a currently-emerging coating technology which makes it possible to use an electric current to vary light and heat transmission through windows.

Task/Ambient Lighting

To insure maximum personnel productivity, lighting must provide not only the correct amount of illumination but good quality in terms of minimizing direct glare, reflected light and lighting contrasts. This requirement is accentuated in the intelligent building because of the accelerating use of workstations with video display screens.

Consisting of vertical rather than horizontal work surfaces, these screens are internally illuminated so the need for ambient or ceiling lighting is minimized. In fact, high levels of ambient lighting can and, in many cases do, cause visual problems due to direct glare, light reflected from screens, and uncomfortable contrasts between screen and background illumination levels.

Consequently, there is a marked trend in intelligent buildings towards task/ambient rather than uniform ceiling lighting. Capable of being varied in terms of intensity and color based on specific job requirements, task lighting consists of either specially-located, fluorescent fixtures with low-brightness lenses and directional louvers or desk-type lamps that are built into workstations and office furniture. The lamps may be a new type of fluorescent tube which is much shorter than conventional tubes -- only 9 inches long -- and can be plugged into power outlets.

Ambient lighting can be provided in a number of different ways. One is in the form of daylighting. Another is indirect lighting produced by floor-mounted, high-intensity discharge lights or other types of lamps which can be used to reflect light off ceilings. In addition, uniformly-spaced, pendant-mounted, surface-mounted or recessed-ceiling lighting can be used.

Task/ambient lighting provides high-quality illumination that eliminates undesirable glare and reflection problems on video display screens. The lighting also conserves energy and can be installed at a lower capital cost in new intelligent buildings.

Higher-Efficiency Lamps, Devices

A number of higher-efficiency lamps which produce more lumens per watt are now becoming available, including argon-filled tubes and high-intensity discharge and sodium lights. New electronic ballasts efficiently provide the same amount of light with 10 to 20 fewer watts. Programmable ballasts make it possible to use electric current to vary light output.

SOLVE ACOUSTIC PROBLEMS

An intelligent building may have greater acoustic problems than other types of buildings because of the growing use of workstations which require the operation of impact printers.

These problems can be solved through the use of vertical, movable acoustic baffles which are hung above workstations. Four feet long, one foot high and one inch thick, these baffles can be moved in any direction to absorb printer noises. Acoustically-treated furniture and stub partitions are effective but care must be exercised to assure proper air circulation through workstations.

ACHIEVE TOTAL BUILDING PERFORMANCE

The advent of the intelligent building is spurring a total re-evaluation of how buildings are designed, constructed and operated or managed from initial concept throughout useful life.

Buildings can offer the ultimate in high tech services, but fail to respond to individual occupant needs for adequate flexibility, power, health and comfort. Consequently, the building can no longer be viewed as a passive setting for office work. It is now, in fact, a holistic system critical to the operational and economic performance of its occupants.

Based on serving people rather than creating technology for its own sake, this development of intelligent buildings from this overall point of view will make it possible to maximize total building performance in terms of HVAC system, lighting, acoustical and energy management capabilities.

Chapter 11

Planning For The Intelligent Building Of The Future

Carl F. Klein
Johnson Controls, Inc.

What Is It?

To plan for the "intelligent building" of the future, we must first define this new buzz word. The intelligent building of the future is not just a building with the latest building automation system in it. It is a building that integrates computers, communications, and office automation with building automation to organize a building's support functions into one package for the owner. Functional areas, such as communications and office automation, are integrated with building automation and control to reduce a building's operating costs for the owner and tenants. An intelligent building, therefore, uses the latest in computer and communication technology to create a highly productive working environment that can be shared by both building users and operators to accomplish their work more efficiently and comfortably.

The key point to this definition is integration. Integration permits each tenant to have access to the comfort, safety, and security provided by a Building Automation System (BAS), the services available with office automation, and the communication capability available with a multi-dimensional telecommunications system that provides not only voice but also data, teleconferencing, telex, and facsimile communications. Figure 11-1 shows some of the integrated building communication and automation services that would be accessible to building users and operators.

Fig. 11-1 Integrated Building Services Model

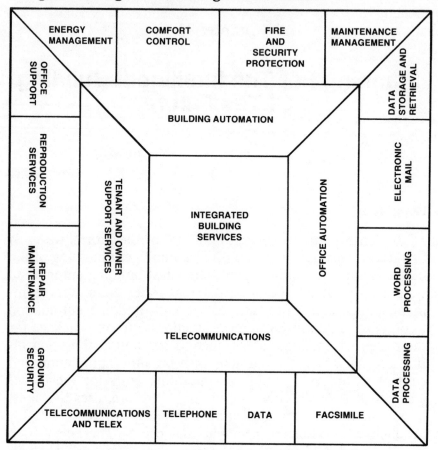

The intelligent building responds to both the needs of the tenants and of building owners and operators. It does this through an integrated communications network and an automated building systems and services program that is designed into the building to meet its present and future needs. Possible synergistic benefits from an integrated PBX/BAS (Private Branch Exchange/Building Automation System) are shown in Figure 11-2. Here we can see that voice I/O (Input/Output) will have important building control, energy management, fire and security applications, since it will, in essence, make every phone an operator interface.

Phones will also be able to act as field processing units for the BAS since they will contain thermostats, humidistats, and other sensors for monitoring the building's environment for tenant comfort and safety. Weather station data will be used to provide anticipatory comfort control and energy savings, and data processing and word processing software will be available for processing BAS data and reports. Energy accounting by person, department, and building will also be possible.

In the area of fire safety, there will be numerous applications where a voice synthesis module might be tied into the system (over a phone line) to notify building occupants of problems.

In the security area watchtour I/O through phone stations will be possible and phones connected to voice print or tone-in voice-out equipment will be available as alternatives to card access systems.

Planning For The Intelligent Building

Planning for intelligent buildings of the future requires a delicate balance between sticking to a long-term strategy and being flexible enough to seize opportunities as they arise. Since significant events that were not expected usually occur during the planning process, success in planning often comes from investigating areas that traditionally would be overlooked.

Fig. 11-2 PBX/BAS Synergy Benefits

I. Energy Management

    A. Voice I/O for BAS; a "Speak and Tell"
    B. Every phone a thermostat, humidistat, etc.
    C. Lighting control - no wall switches
    D. Weather station data
    E. Operations flexibility over space and time
    F. Word processing for BAS reports
    G. Energy accounting by person/department

II. Fire Safety

    A. Voice I/O for alarms
    B. Every phone a pull station/smoke or heat detector
    C. Evacuation control

III. Security

    A. Watchtour I/O through a phone station; dynamic watchtours
    B. Security alarming and logging through a phone
    C. Infra-red person detectors - in a phone
    D. Phones replace cards for access control stations

This process will be true when planning for intelligent buildings. Ten years from now, when intelligent building technology is in place, it might not be all that is currently expected. Rather, it might be paying off in ways that at present are unpredictable. Planners can benefit from this knowledge, however, only if they take advantage of unexpected opportunities and develop a plan that anticipates and manages change.

Intelligent building planning begins with a forecast. For a developer of intelligent building systems, this forecast anticipates technological changes and assesses their market value. In other words, it asks:

- What problems will need to be solved.

- How many customers will want to solve these problems during the coming weeks, months, and years.

The intelligent building developer then assesses the capabilities required to economically solve these problems and works out, through either development or acquisition, when those capabilities will be available so solutions can be brought to the market.

For building contractors who know the product they want to bring to the market, the process is different. Their questioning sequence is: Where is the system we want to supply the building owner? What problems will we have in making it (putting it together)? Where can we find solutions?

Success for both the developer and contractor comes when the right solution meets the right problem at the right time. This is fairly easy when the technology customer is looking for incremental solutions or small improvements. The uncertainty, however, increases significantly with an objective such as the functional and structural implementation of the intelligent building of the future.

Changing Business Environment

Planning for the intelligent building of the future requires a look at the entire integrated intelligent building business. The

BAS/HVAC (heating, ventilating and air conditioning) business appears to be broken down into the following categories:

1. BAS/HVAC Equipment Manufacturers
- BAS only

2. BAS/HVAC Suppliers

3. BAS/HVAC Installation

4. BAS/HVAC Service

5. BAS/HVAC Building Operations

The building communications marketplace or selling environment appears to be broken down into similar categories:

1. BCN (Building Communications Network) Equipment Manufacturers

2. BCN Equipment Suppliers

3. BCN Service

4. BCN Operations

From this component breakdown of the BCN and BAS/HVAC businesses, the significant similarities between the two business environments can be seen.

Building automation and communications can be viewed as being in a changing or transitional business/market. The convergence of the computer communications and building automation and control technologies is producing a redistribution of functionality in buildings as shown in Figure 11-3.

This redistribution is taking place in two phases: The first phase is structural and the second phase will be operational. The structural phase consists of using shared communication wire and hardware, while the second phase will consist of performing expanded operational functions to overlap those performed in other networks.

Up until the last several years, a variety of local building communications systems were installed side by side for building automation, telephone, office automation, computer data communications, etc. The proliferation of communication networks and their overlapping functionality are making it desirable to combine several of these networks.

The function of VLSI (Very Large Scale Integration) technology and its rapid development and expansion into the area of telecommunications is making the development of more general purpose communication networks possible. As these networks evolve, they will be able to meet the needs of a broader range of functional requirements as shown in Figure 11-1.

Building owners and real estate developers are currently considering implementing a variety of shared telecommunications and computer-based services to their tenants, such as phone service, energy management, integrated office and security services. This new business area is being called the joint tenant market.

Significant Characteristics

The three most significant characteristics of the intelligent building of the future are efficiency, flexibility, and effectiveness.

Efficiency in intelligent building terminology means controlling operating cost. Here building comfort must use no more energy than is absolutely necessary and communications are implemented in the lowest-cost mode.

In intelligent buildings, building automation, data processing, and telecommunications equipment work together to help increase the output of staff personnel and hold down rising labor costs. Intelligent buildings should therefore provide a more productive environment for the knowledge worker of the future. Some developers envision intelligent buildings with teleports--transmission and receiving centers--that link a corporation's offices nationwide with enviable efficiency. An integrated data processing, telecommunications and building automation network within a building will give tenants access to electronic mail, voice and data services, and least-cost rout-

Fig. 11-3 Redistribution Of Functionalty In Buildings

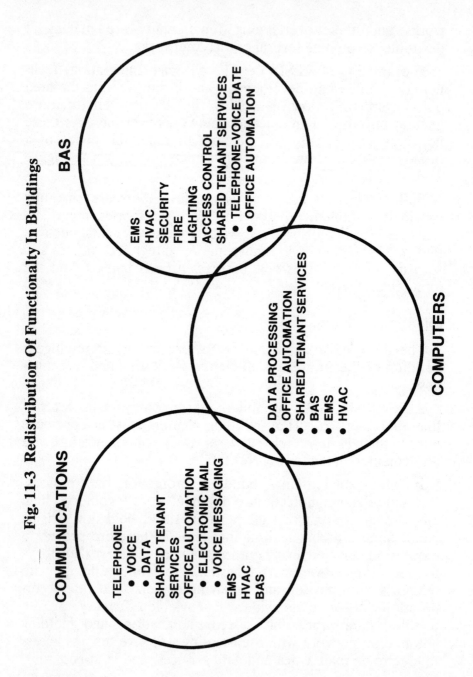

ing for telephone calls, while permitting owners to slash energy costs for lighting and temperature control.

Flexibility keeps intelligent building equipment economical in the face of change. Flexible equipment is equipment that is:

1. Reprogrammable

2. Functionally distributed

Flexibility also means the use of intelligent wiring. Intelligent wiring consists of a network capable of supporting a multiplicity of functions, such as telecommunications and data processing for office and building automation. The key to intelligent wiring will be the digital PBX capable of supporting both voice and data information simultaneously.

Flexibility will also permit easy functional and structural expansion to accommodate future growth. For example, fire safety and security features can be added at the same time money is saved through energy management.

Effectiveness in the context of intelligent buildings means knowing the right things to do when managing a building--things such as control, energy management, maintenance management, self-diagnosis, communications management, fire and security protection, and office services.

More advanced controls permit buildings to anticipate and compensate for changes in weather conditions and occupant use. The energy management systems monitor any combination of boilers, fans, pumps, air handlers, chillers, and lighting and optimize their most efficient use of energy. Tenants are able to lease computer hardware and software for applications like word processing and financial analysis. They are able to pick and choose from a menu of communication functions including telex, electronic mail, information retrieval systems, and facsimile transmissions in addition to voice. In the area of fire safety, sensors not only ring an alarm bell but call the fire department, control the elevators, pressurize the floors above and below so the smoke can't spread, and exhaust the smoke on the floor where the fire is taking place.

Success in the intelligent building business will center around how companies respond to change. They have a choice of either managing change or adapting to change. The difference is that in managing change companies decide what the future business should be like and then take steps to make it that way. This approach requires careful thought and strategic planning. Adapting to change, on the other hand, requires no plan at all. This approach seldom brings long-term success. A company simply needs to follow the flow. Obviously, managing change and then planning is the way to make continuous new contributions to business growth.

SECTION III

BUILDING WIRING --
STRUCTURAL INTEGRATION

Chapter 12

Transmission Media: Wiring For Building Intelligence

Rod Elger
Johnson Controls, Inc.

Introduction

The growing variety of transmission media requires that the characteristics, advantages, disadvantages and suitability of each be understood. Transmission media applications in local area networks (LAN's) are based on many factors including cost, availability, and practicality. Emphasizing LAN transmission media which are specified by the IEEE 802 Committee (i.e., twisted-pair, coaxial cable, and fiber optic cable.) This chapter discusses those factors that distinguish media performance and applicability. First transmission media and their relation to LAN's are discussed. Applications of transmission media are then reviewed, followed by a description of the kinds of media and their functions in building automation systems (BAS).

A) Transmission Media Definition

Communications between various stations in distributed processing networks is achieved by means of transmission devices and media. Transmission media provide the physical channels used to interconnect nodes in a network. Any configuration of transmission facilities existing between any two communicating stations may be called communications media. Media are classified as bounded, e.g., wires, cables, and optical fibers, or unbounded, e.g., "air waves" over which radio, microwave, infrared, and other signals are broadcast.

B) Media To Network Relationship

In recent years, there has been increasing interest in methods of optimizing the data acquisition and control functions of automated building systems. System requirements have become more sophisticated due to the growing number of controls, stiffening environmental constraints, and escalating labor costs of skilled building operators and personnel involved in installation and maintenance.

The choice of transmission media is significant because the cost and labor involved in its installation and maintenance is a major expense. It is important that a bid for either a new or existing building offer the best price. The cost advantages of alternate media must be understood well enough at the time of bid to incorporate them. Although there are exceptions for technical merits, most contractor selection is based on bid price.

The key to optimizing a diversity of products lies in planning, designing, and installing the products based on specific applications. Determining which communication medium is best for a particular application in an automated building depends on the environment and communication requirements specified by the user. There is no effortless method of media selection; it requires choosing the best medium based on current and projected needs and currently available information.

The most successful networks represent not a single but many technologies. There is no single network design or transmission medium that will solve the overall communications problem. An open architecture is needed that will provide a symbiosis of topologies, media, and communication technologies to accommodate the cost and performance requirements dictated by particular applications. Mixing network and computer technologies is the trend with the combination dependent on the applications.

Overall, in the 1980's, it is expected that the technological emphasis will be on achieving greater integration between building system monitoring and control and office automation, and obtaining substantial economic gains by carefully customizing communications facilities to the unique characteristics of customer building application environments.

C) Transmission Media Viewpoints

Communications functions, as viewed from outside the communications system, are usually called services because the results of corresponding control operations take the form of services offered to users. Users of these services do not have to know what communications functions are being used. Telephone users, for example, want to talk. As long as a network fulfills this need, they do not preoccupy themselves with whether their voices are going analog, digital, fiber optic or microwave, or what protocol, line conditioning, or modulation technique is used. Ultimately, it does not matter to users how the data gets there but only that it does.

However, the situation is different when users want to operate a communications subnetwork such as a BAS in a larger system which also performs communications functions. Then, it is not enough for users to know only values of performance criteria. They must also know the communications function methods used in order to design adequate interfaces. Usually, users will expect vendors of BAS subnetworks, to satisfy their demands for cost effective, reliable, flexible, interfacing. Designing LAN's that satisfy users is not a simple process since it requires choosing transmission media, accessing scheme network topology and other elements.

D) Design Elements Of Local Area Networks

The design elements of LAN systems must be carefully chosen to provide required levels of data communications capabilities and network performance. The choices made determine the very nature of a network. Design elements for LAN's can be grouped in the following categories:

1) Topology

2) Control, Access, And Allocation Of The Network Channel

3) Switching Techniques

4) Transmission Media

The most important point about these classifications is that they are largely independent of one another. In theory, any control scheme can be used in conjunction with any access mechanism on any medium in any topological arrangement using any of a number of switching techniques. For example, LAN's can be configured in tree networks with either dedicated, polled, or random access procedures, and any combination of these choices can be operated over twisted-pair, coaxial, or fiber optic cables. Coaxial cable, inturn, can be either baseband or broadband. In addition, hybrids are frequently created within categories. For example, a network can combine elements of several different topologies, access mechanisms, or transmission media.

But, although these various combinations are possible, some operational groupings are poor or even absurd from a cost or performance standpoint. For instance, it is not economically feasible to use a broadband coaxial cable to accommodate a few teletype speed devices. Alternately, transmitting video over a twisted-pair, random access network would inevitably lead to disastrous performance.

The entire paper, "Research Report On Transmission Media: Wiring For Building Intelligence," by Rod Elger, Research Engineer, has been published by Johnson Controls. The paper gives guidance to consultants, developers, building owners, and tenants interested in wiring options.

The paper discusses those factors that distinquish media performance and applicability. Transmission media and their relation to LAN's are described. Applications of transmission media are then reviewed, followed by a description of the kinds of media and their functions in building automation systems.

APPLICATIONS OF TRANSMISSION MEDIA

A) The Media Independence Concept

The IEEE 802 Committee is working toward the development of standards for so-called physical and link levels for LAN's. In these two lower layers are specifications for the physical transmission medium and the way nodes should interface with this medium. The committee has conceptually determined that standards should as much as possible be independent of transmission line types and signaling techniques, i.e., the standards should be media-independent. More and more systems will become available in the future, permitting users to choose media that serve them best. Media independence can be obtained by matching the special capabilities of each medium with the different requirements within a network. In effect, users have the freedom to choose the medium or combination of media that will provide the capabilities required without paying for more than needed. With multimedia transparency, the option remains open to use other media in the future without investing more now.

B) The Media Access Unit

To provide a standard for today's technology and yet keep open the possibility of other signal-transmission technologies in the future, the IEEE 802 standard defines a box called a Media Access Unit (MAU). The MAU allows the use of conventional copper cable, yet can also easily accommodate optical fibers. To make a standard for the physical layer, a transmission line with its connectors must be selected and electrical signal levels defined.

To obtain this flexibility for transmission media, one side of the MAU will have a fixed standard interface that connects to particular equipment such as a computer or terminal. The other side will provide an interface to a particular type of transmission line. Three transmission-line-interface techniques are currently being standardized:

- A coaxial cable with baseband data signaling (no carrier for the data).

- A coaxial cable, commonly used for cable television with data signaling performed by RF carrier.

- Fiber optics, which shows great promise and is in the preliminary stage of standardization.

The actual physical connections from a Communications Interface Unit (CIU) to the medium are made in a variety of ways by Media Access Units (MAU's). Most often, the MAU is a form of cable tap from which a wire is run to the CIU. Standard multipin connectors, such as those conforming to RS 232-C or RS 499 specifications are also used.

With either method, the data is sent and received serially. The MAU merely contains the necessary line drivers and receivers tailored to the particular transmission line. If a superior transmission technique is developed, only the MAU will become obsolete rather than all the equipment on the network. The cable in the MAU interface can be up to 50 meters long. Standard data rates for the network will be 1, 5, 10, and 20 M bps. Most of the physical layer's functions will be carried out on the equipment side of the MAU interface. These functions include data coding and decoding synchronization and recognition of the start and stop of a data unit.

Choice of transmission media for data communications brings up the simple question: "Should a single medium be used, or is there an advantage in the mixed-media approach?" The concept of building a data communications facility with a single medium is certainly the simplest approach and is good when traffic is anticipated to be uniform. On the other hand, this approach may be mismatched to certain traffic types flowing over the system. It also may be vulnerable to various kinds of errors or damage and inflexible when present traffic levels grow or the nature of the traffic changes. A mixed-media system is potentially suitable for more traffic types and certainly more suitable for heterogeneous traffic and applications in which growth and change are anticipated. Successful special purpose LAN communications systems will have to combine various communication media technologies to accommodate the cost and performance requirements dictated by various building automation applications.

C) Multimedia

Successful general-purpose LAN communications systems must use the most appropriate communications medium for any specific application.

Experience indicates that there is no one ideal solution to each and every problem encountered in configuring a data network. When a search is made for the most capable medium for long-term installation growth in a multimode local environment, it must begin by comparing the characteristics of available means of transmission.

Several types of data media can be combined into a trunk wiring system to develop a hybrid system. For example, a series of small motors within a system can be turned on or off via an FM radio signal. In another area, a coaxial cable can be used to communicate with remote buildings while a more expensive coaxial cable can be used for buildings which are closer. In such arrangements, each aspect of a system must be specified precisely to insure proper control and maintenance.

D) Media Selection Criteria

Although the transmission medium is not considered part of the OSI reference model, it is one of the key attributes by which different physical layer implementations in common use today may be organized. It is axiomatic that the most important element of a data communication system is the channel. Its bandwidth limits the volume of information that may be transmitted within certain time limits. Any imperfections such as interference and distortions that may be present also affect the maximum possible volume of information and its accuracy. In addition, these factors contribute to the complexity of terminal devices.

Transmission media used for local networking differ in raw transmission capacity (bandwidth), potential for connectivity in terms of point-to-point or broadcast capabilities, geographic scope allowed due to attenuation characteristics of the medium, immunity to noise, relative cost in terms of the actual hardware purchase required and its installation, and suitable applications. (The cost of labor involved in the installation and maintenance of the medium is the overriding expense.)

The choice of transmission media for a system involves the analysis of various transmission paths which could be used to support communications. It may also involve choices among types of uses of the media and methods of obtaining communications services. The following section describes criteria that are useful in comparing relative performance of various types of transmission media:

1) CAPACITY describes the available medium bandwidth and ease of establishing independent logical channels for each network service type. Bandwidth is a convenient measure of signal speed useful for media comparison. Similarly, the rate at which a system can change stored energy is reflected by its useable frequency response measured in terms of system bandwidth. Speed is an important property of any network. Efficient network utilization requires minimization of transmission time, i.e., sending the most information in the least amount of time. Rapid information transmission is achieved by using signals that change rapidly with time. However, signaling speed cannot be arbitrarily increased, because the system will cease to respond to signal changes. Various transmission media are capable of differing information rates. But transmitting a large amount of information in a small amount of time requires wideband signals to represent the information and wideband systems to accommodate the signals. Bandwidth, therefore, emerges as a fundamental limitation. Generally speaking, the higher the network bandwidth, the more exotic the transmission technology, the more expensive the physical medium, and the more restricted the ways in which the network can be configured.

The long-term rate at which data can be exchanged between a pair of devices depends on the amount of competition for channel capacity, the control system, and the capacity of the interface between user equipment and network.

2) GEOGRAPHIC SCOPE refers to the maximum distance between points on a network. Power is attenuated due to dissipative or radiative effects as it travels further from its source. With more power, information can be transmitted longer distances. Geographic scope is therefore affected by the media capacity for transmission power with and without repeaters/amplifiers, and by the amount of inherent attenuation.

3) EXTENDABILITY measures the network's ability to expand geographically with minimum impact on an existing cable configuration.

4) FLEXIBILITY describes a network's ability to accommodate both new services and upgrades or expansion of existing services. Provisions for growth affect several elements of the system. The most critical items are the CPU, its memory, and its software programs. Field wiring and Field Interface Devices can also be designed to minimize the impact of expansion of all known future points and new services.

5) CONNECTIVITY encompasses a network device's ability to access other devices in a straightforward manner. Certain transmission media are capable of point-to-point and/or broadcast connectivity due to the nature of their construction and implementation. Direct wire connections may be more applicable for point-to-point communications, while radio may be best suited for broadcast situations. Included here are limitations for multidrops from a main segment of media where applicable with considerations for repeaters and/or amplifiers.

6) CONFIGURATION includes the medium's ability to simply and reliably service each user's point of access. This design consideration involves the inherent relationship between network media and topology.

7) MAINTAINABILITY is the ability to maintain a system which is extremely important. One such concern is the ability to support field maintenance through training programs and manuals. Equally important, if not even more so, is the availability of professionally-trained maintenance persons employed, licensed, or authorized by a BAS vendor. Maintainability of media should be considered in design. While a high level of sophistication may be desirable, it should be recognized that increased sophistication often results in increased maintenance problems and costs. As such, trade-offs may be necessary.

8) RELIABILITY relates to two issues:

How well a system performs. If media is subject to frequent breakdowns, it has a low level of reliability.

- The way in which a system performs. If it has been designed and installed well, it should function well so those who rely upon it have confidence in it. On the other hand, the system may be designed in such a way that there are certain built- in problems. For example, occasional interference with transmission lines may result in signals that are improperly received, lost, or erroneous.

Major factors relating to reliability include:

- The average length of time for which a system performs completely as it should without breakdown.

- The average extent of downtime.

- The nature of problems experienced and the degree to which they inhibit a facility from achieving its mission.

9) NOISE is a contamination to a network that alters the signal shape, causing an unintended signal perturbation. It is a broad classification that actually includes three effects: distortion, interference, and pure noise. Distortion is signal alteration caused by an imperfect response of a system to the desired signal itself. It disappears when the signal is turned

off. Interference is contamination by extraneous signals, usually man-made, of a form similar to the desired signal. Pure noise is the random and unpredictable electric signals from natural causes that are both internal and external to the transmission system. When such random variations are added to an information-bearing signal, the information may be partially masked or totally obliterated. Pure noise cannot be completely eliminated even in theory. References to noise means pure noise.

Typical noise variations are quite small on the order of microvolts. If signal variations are substantially greater, then noise may be all but ignored. The signal-to-noise ratio, measured in decibels (dB), is therefore sometimes large enough for noise to go unnoticed. But this is not always the case. Differing transmission media have varying amounts of immunity to noise due to their construction and nature of operation.

Based on the above characteristics, each type of media is best suited to certain generic types of applications. For example, the broadcast nature of radio implies a quite different set of applications than the point-to-point nature of twisted-pair.

Transmission media also differ in relative cost due to the type of equipment required for information transmission, transportation, and reception and installation, maintenance and life-cycle costs.

Information transmitted in a local network may travel through wire, cable, radio waves, or light beams. Each type of channel produces certain constraints on factors important to a BAS manager. These factors include the amount of information that may be transmitted in a given period of time, the quality of information when received, the size of a network and its cost. Therefore, for each type of media, there follows a general discussion of construction and technology along with details relating to the areas of comparison.

KINDS OF MEDIA

A) Bounded

1) TWISTED-PAIR WIRE   Twisted-pair wire was one of the original wire types used in telephone communications and it remains the main form of media in place for local telephone and data transmissions. Pairs of wires are twisted together in a regular geometric pattern in order to make electrical properties constant throughout the length of the line and minimize the interference created when adjacent pairs of wire are combined in multipair cables. Radiation from the twisted-pair can occur when the relationship between conductor separation and operating frequency reaches a certain point. Consequently, definite limitations on frequency of transmission exist.

Transmission over twisted-pair can be either analog or digital, using a variety of signaling approaches. Digital techniques often make use of Pulse-Code Modulation (PCM).

Depending upon distance, signaling techniques, and quality of wire pair, information can usually be transmitted at several hundred kilobits/second and even faster in point-to-point situations. If repeaters are spaced at sufficiently close intervals, speeds in the magabit/sec range can be achieved. In PABX systems, typical transmission lines operate at 64K bits/second maximum with user device lines operating at 9600 bits/second. Multipoint applications are limited to fairly low speeds (1200 bits/second) due to load and capacity considerations.

Twisted-pair can be employed for both point-to-point and multidrop applications. When used for multipoint situations, however, average device data rates are severely restricted, decreasing as the distance increases between devices due to propagation delays. Device burst rates may still be high but a large number of devices will restrict concurrent usage for all devices. Point-to-point is far more common and most suitable for general local network application.

Twisted-pair can range over an area of fifteen kilometers, provided that suitable conditioning takes place throughout the length of the line. Most implementations utilizing twisted-pair restrict distances to within buildings. Without a modem, the use of a pair of wires is restricted either to low speeds or to short distances.

Energy loss is an important parameter to consider when discussing geographic scope and network security. As distance between communicating devices increases, energy loss will also increase consequently, at some point, a receiver will not be able to properly respond to and detect incoming information. The lost energy radiates outside transmission lines, allowing motivated outsiders to detect and pick up the energy. There are two ways in which energy, impressed at the sending end of a transmission line, may become dissipated before reaching its destination: radiation and conductor heating. Radiation losses arise because a transmission line may act as an antenna if the conductor is an appreciable fraction of the transmitted wavelength. Such losses are difficult to estimate since they normally are measured rather than calculated. Conductor heating is proportional to current and impedance, increasing with frequency as the signal is carried more toward the outside of the conductor.

Twisted-pairs of wire can be purchased with a variety of properties at a variety of costs. They are available with a varying number of twists per foot, with no electrostatic shielding, with braided shields, or with solid shields. In general, good noise immunity can be achieved with effective wavelengths much longer than the "twist length" of the cable. In balanced low-frequency systems, noise immunity can be as high or higher than for coaxial cable. However, at frequencies above 10 to 100 KHz, coaxial cable is typically superior.

The high shunt capacitance of twisted-pair can cause signal distortions. Quality of analog signal transmission can be improved by utilizing devices known as loading coils. These inductive components offset the shunt capacitance, which builds up as the length of the communications line increases. The loading coils tend to enhance voice transmission quality within the 0 to 4 KHz voice band, but are not especially useful if a line is to be employed for wideband data transmission. The point is that an extremely wide variation in noise immunity and other properties for twisted-pair products and for networks utilizing those products are available. Therefore, close examination of requirements versus product limitations should be made.

The twisted-pair medium is used for a variety of communication applications. But its connectivity and noise immunity limitations must be recognized. Unless properly protected, it should remain within buildings or rooms. When part of a bus network, for example, typical applications involve interfacing devices with the network rather than serving as the general network medium. Twisted-pair is useful in multipoint applications when very low speed, low-duty cycle devices are to be interconnected.

Twisted-pair is the least expensive of the transmission media in terms of cost per foot. However, considering the amount of conditioning that may be required and the medium's connectivity limitations, installation costs may approach other media such as coaxial cable.

No major bandwidth technology changes are expected to make twisted-pair wires competitive with coaxial or fiber optic cable. Bandwidth has been a fundamental problem, but it is now known how to drive conditioned twisted-pair well over a megabit per second. The revitalization of twisted-pair is coming from the needs of personal computer uses due to its lower cost. In addition, bandwidth is quite adequate for this application. As personal computer local networks proliferate, this media will continue to find application.

The nominal bandwidth of unconditioned twisted-pairs is between 300 and 3000 Hz. For each Hz of available bandwidth, 2 bps may be transmitted. Data transmission in twisted-pairs, in most cases, is limited to 1200 bps or less. Hardwired twisted-pairs must be conditioned in order to obtain operating speeds up to 9600 bps.

Data transmission between Field Interface Device (FID) and Multiplex (MUX) panels can use line drivers operating at a speed selected by the system supplier.

Because of their configuration, twisted-pairs are difficut to terminate. Fully automatic procedures are not yet available and manual methods are slow.

2) <u>COAXIAL CABLE</u> Coaxial cable is a form of transmission line very similar in concept to twisted-pair but with modified construction to provide different operating characteristics. Coaxial cable is one of the most commonly used data transmission media. Coaxial cable currently used for local area networking is generally classified in two ways according to modulation techniques employed: baseband and broadband. There is a slight physical difference between the two types of cable. Baseband coaxial cable has a center conductor surrounded by a dielectric insulator. Surrounding these two is a woven copper mesh. A polyvinyl jacket or other coating is usually then placed over the mesh to protect against weather and handling. Broadband coax, like baseband, has a center conductor surrounded by a dielectric insulator. Instead of a copper mesh, however, an extruded-aluminum jacket is used. The coaxial cable typically used for baseband and broadband signaling in LAN's are 3/8 inch and 1/2 inch, respectively. While each of these provides different amounts of bandwidth and performance characteristics, the substantial difference between the two is the way they are used. The capacity of the coaxial cable is used in baseband LAN's primarily to transmit a single baseband signal at very high data rates, in the area of 12 M bits/sec.

In broadband LAN's, the capacity is used to create a large number of frequency subchannels from the one physical channel. Coaxial cable commonly used for Community Antenna TeleVision (CATV) has available bandwidths in the range of 300-400 MHz. This translates into enough capacity to carry over 50 standard six MHz color TV channels or thousands of voice-grade and low-speed data signals (e.g. 9.6, 19.2, or 56 K bits/sec.).

The widespread use of coaxial cable is due to its moderate cost and ready availability. In addition, the technology for installation, connection, and transmission control on coaxial cable is well-developed. A variety of taps, controllers, splitters, couplers, and repeaters are available that enable the cable to be easily extended and branched off to reach user locations for connection of user devices.

Coaxial cable has also become a popular medium in LAN's because of its large capacity, low error rates, and configuration flexibility. It is an excellent replacement for twisted-pair wire in meeting the high-performance criteria of local communications, e.g., point-to-point links between devices in star or ring configurations and for multidropped bus configurations. Its multiplexing capability means that there is no practical limit on the number of facilities which can be connected to the system, making it an excellent choice, especially when expandability is a concern.

Baseband coaxial cable is usually of 50 ohm grades, capable of sending a single signal using digital techniques. Connection to the cable is often through a nondestructive tap with a passive transceiver. Highly reliable repeaters, amplifiers, and taps are available. Considering that transmission is without modulation (relatively low transmission frequency), construction issues of baseband coaxial cable (e.g. relationship of conductor to transmitted wavelenghts) may present security problems, in that the cable may act as an antenna allowing eavesdroppers to tap into the line with appropriately placed pickup coils.

Broadband media concepts are implemented primarily with off-the-shelf CATV (Community Antenna TeleVision) hardware, generally using 75 ohm cable for dual and mid-split systems. Low cost signal splitters and taps achieve branching of cables and commercial repeaters (line missing between failures for these CATV devices has been established at 400,000 hours or better in large installations.) Fully redundant repeaters are also available and status monitoring equipment for these units has recently been introduced. All equipment is mass-produced and extremely reliable due to heavy use by the CATV industry.

Baseband transmission implies no modulation of the digital signals transmitted; transceivers drive data onto the cable using a variety of coding techniques such as Manchester phase encoding. Information is transmitted in bit serial fashion; one signal occupies the cable bandwidth.

Networks using broadband coaxial cable often employ frequency and phase modulation techniques, transmitting analog signals along the cable. In its most simple form, a single broadband cable may be characterized as a two-way radio frequency medium for subsplit systems with a forward bandwidth of 140 MHz and return bandwidth of 105 MHz. Signals are transmitted on a return spectrum channel and retransmitted at a higher frequency on a related forward channel. All points attached to the broadband coaxial cable can receive the retransmitted signal.

Bandwidth differs significantly according to the mode of transmission. Existing baseband coaxial network implementations are limited to a single signal with backbone cable rates between 3 and 10 M bits/second. Such limitations are not necessarily a result of cable technical properties but of total network cost with manufacturers making the resulting tradeoffs. High-speed multipoint implementations do exist (50 M bits/second) with specialized high-cost hardware.

Broadband coaxial cable has a capacity that is midrange between fiber optics and baseband cable. For each trunk (main cable segment) an ultimate of approximately 150 M bits/second of full-duplex transmission path is available for a total of 300 M bits/second. All channel assignments are achieved by frequency- division multiplexing in which a channel may be as wide as the full capacity of the cable. Since network manufacturers build systems that can share cable with existing operations, such as CATV (67 MHz), bandwidth may be cut accordingly.

Coaxial cable is applicable to point-to-point and in, particular, to multipoint topologies. Conventional topologies may be implemented (star, ring, bus, mesh), and complexity can range from simple to sophisticated. Considering the variety of access and multiplexing techniques available, variations of the bus topology are particularly applicable. Systems can very often be implemented using only one cable.

Broadband cable also has greater immunity to electro-magnetic interference than baseband. Connection is usually accomplished by splitters, which are widely used in the cable television industry. Adding new devices to broadband is not as simple as with baseband since the cable must sometimes be cut and a splitter spliced in. Also, tuning of the network for impedance and signal level must also be considered.

Multidrop capabilities of a single cable segment for the two types of cable are highly dependent on applications and desired data rates. Baseband multidrops are on the order of 100, while broadband can support several thousand. The basic difference consist of whether single or multiple signals can be transmitted simultaneously. Repeaters and/or amplifiers must be placed approximately every 600 to 1600 meters to respectively regenerate signal shape or amplify signal power to original levels.

Depending on tolerable delay, load, and implementation, maximum distances in typical baseband coaxial networks are limited to one to three kilometers; broadband networks can span areas of ten kilometers or more (50 is a practical upper limit).

The basic difference between the two distances lies in technical considerations based on information transmission mechanism (analog or digital), modulation techniques, transmission frequency, and attenuation properties of the two types of cable. The types of electromagnetic noise usually encountered in industrial and urban areas are of relatively low frequency. Therefore, information transmitted on baseband in digital form (square waves) may be highly susceptible to this noise. Repeaters placed along the transmission path may employ filters to remove noise but limitations on distance still exist. Analog information modulated on a carrier is less susceptible to the types of noise frequently encountered. Modulation techniques that reduce noise effects coupled with higher frequency transmission (low frequency noise would not hamper reception) allow a wider geographic scope for broadband. Inherent attentuation of the types of cable utilized must also be considered.

Immunity to noise for coaxial cable networks is highly dependent upon their application and implementation as described in previous sections. Typically, baseband systems have an immunity of 50-60 dB in isolation while broadband has a comparable figure of 85-100 dB.

Coaxial cable networks are particularly applicable to sub/tree architectures, such as those that might be useful for office automation, laboratory, and process control environments. Devices requiring network communication may be scattered throughout an industrial complex, tapping into appropriately placed cable segments. Depending on implementations, devices may communicate at a variety of data rates, and hence may range from low-speed interactive terminal links at 300 bits/second to high-speed computer to computer links (2-5 M bits/second). Broadband coax has a large capacity supported by high multidrop capability, applicable to data distributions networking. In addition, the cable can be shared with other networks or systems.

Per-foot cost of coaxial cable is not high. It is currently more expensive than twisted-pair but less costly than fiber optics. Depending on the application, however, total installation cost may actually be lower than for twisted-pair, considering the multipoint capabilities of coax versus the latter's point-to-point limitations.

Due to their differing properties and capacities, the cost of the broadband coax is approximately 1-1/2 times that of baseband. Office automation environments tend to require fairly low-cost services, making coaxial cable quite viable.

It is expected that more refinements will take place in broadband cable and its associated devices than in baseband due to the significantly greater capabilities of broadband.

3) <u>FIBER OPTIC CABLE</u> Optical fibers are currently used only in a limited range of applications. Transmitter, receiver, and connector technologies in fiber optics are considerably behind those of coaxial cable and twisted-pair as are multiplexing capabilities.

Fiber is most frequently used as a high-performance medium for half-duplex point-to-point connections. Tapped multipoint connections with dropped nodes are rare since the technology for fiber taps is not well-developed and is very expensive. Because of the dB loss associated with each tap, large-scale multidrop operations requiring tens or even hundreds of connections for a segment of fiber are not feasible without a complex active-tap facility. Experimentally and expensively, the medium can operate in a broadcast environment using optical couplers. Couplers are used to directionally combine energy from one waveguide into two or more. Research has been conducted in the area of multiplexing techniques for fiber optics in which different light frequencies are multiplexed and demultiplexed. But these optical multiplexers are not presently practical. Passive multidrops are currently limited to approximately sixteen nodes. Multidrop network implementations must also recognize the unidirectional characteristics of fiber optics. Two-way communications such as a bus must use two fibers; in ring-type architectures, a single fiber is sufficient.

While not presently practical, a single segment of fiber optics could support many more drops than coaxial cable due to lower power loss at each drop, lower attenuation characteristics and greater bandwidth potential.

The composition of the fiber itself determines the transmission attenuation which is caused either by scattering or absorption of trace elements present in the core.

Fiber optics exhibit losses in the 2-10 dB/Km range. Losses as low as 0.5 dB/Km have been demonstrated under laboratory conditions. Using repeaters and/or amplifiers where required to regenerate digital signals and/or increase source levels can end the fiber optic networks to span a distance of 50 Km or more. Practically, rates between one and 50 M bits/s over a distance of 10 Km are easily achievable. Using mass-produced transmission and reception equipment, it is possible to operate at 140 M bits/second over a distance of 6-8 Km without intermediate repeaters. Cable losses are independent of transmission frequency. Hence, no equalization is necessary.

Based on current costs of creating multidrops from cable, fiber optics is presently most suitable for baseband, high speed, high capacity point-to-point data links between network segments.

An obvious application is a BAS since point-to-point connections are natural to its topology. Examples include:

1. CPU to CPU high speed datalink.

2. CPU to high speed peripheral.

3. Connection between a terminal and a processor over KM range distances.

4. Link between buildings of an industrial complex.

5. Communications path between complexes at opposite ends of a city.

Fiber optics implementations are more expensive than twisted-pair and coaxial cable in terms of cost per foot of cable and required equipment such as transmitters, receivers, connectors. However, the costs are declining as engineering and manufacturing techniques improve. Consequently, it may become a viable alternative for all local network topologies in the future.

Fiber optics as a communications medium is rapidly gaining ground due to progress being made in the key problem areas of:

- splicing
- coupling (multidropping)
- connecting standard agreements on fiber sizes, connector types, topology, wavelength of transmitters, etc.

The physical configuration that is receiving the most attention is the star network. This is because of it's flexibility in being able to accommodate both a logical ring as well as a star wire center by replacing it's core. It is also the most maintenance free topology and the easiest to trouble shoot. The current wavelength being used is 850 nanometers. The reason is cost. When the pricing is right a migration will be made to 1300 nanometers. Connectors are now able to provide less than .2db loss. This means that with a power budget of 20 db one can use 2 to 4 connectors per link. The preferred optical fiber is a fully graded indexed fiber with 100/mb/s data rate, a ring topology, and a token access method, however they have not yet reached final agreement on core size, connectors, transmitter/receiver wavelength and a reconfiguration algorithm. The total system cost for a fiber optic network is now often less than hard wire when one considers:

- installation cost
- the number of repeaters required

It appears fiber optic networking is coming on strong with the only remaining roadblock the agreement on a final standard.

ADVANTAGES

There are many emerging advantages as well as some disadvantages of fiber optics in relation to LAN transmission requirements. The following list of factors starts with more apparent advantages:

1. WIDE SIGNAL BANDWITH - GROWTH

Optical fiber offers extremely high potential data rates or equivalent bandwidths, providing:

- Flexibility in upgrading system capacity without adding new cables.

- Suitability for modern broadband TDM digital communications and PCM data.

- Wide transmission range-bandwidth products (640 MHz -Km) at 44.7 Mb/sec for high resolution color TV signals.

2. COMPLETE ELECTRICAL ISOLATION

- Insensitivity to electromagnetic and radio frequency interference.

- Immunity to ground-loop and cross talk problems. No danger of electrical shorts between conductors or spark or fire hazards.

3. INHERENTLY SECURE

- Fibers do not emit signal and thus thwart security violations such as line tapping with electronic cable.

4. RELIABLE

- Relatively small size, lightweight, high strength, and flexibility.

- Immune to severe temperature fluctuations, rain and snowstorms, bacterial decays, etc.

- Error rates are very low (one bit error per 10^9 bits) with the result that most error detection and retransmission overhead can be eliminated.

5. REGULATIONS

- Fewer government regulatory difficulties due to elimination of frequency allocation.

6. POTENTIAL LOW COST

- High quality, low attenuation fibers are not essential for data processing because of the short lengths involved, resulting in a minimum cost factor.

7. LOW TRANSMISSION LOSS

- 4.4 dB/Km and less to 1 dB/Km between repeaters spaced 4 Km apart.

- Line losses are independent of frequency.

c) DISADVANTAGES OF FIBER OPTIC CABLE

There are some limitations which currently prevent the immediate use of optical systems:

1. Mechanical integrity

2. Detector and emitter response time

3. Termination reliability

4. Cable recycling and repair

5. Moisture absorption

6. Unidirectional

7. Present cost

8. Limited number of taps

9. Lack of standard fiber optic components

Fiber optic terminal equipment selection is limited. There is a lack of skilled installers and maintenance personnel familiar with this medium. Fiber optics must be carefully installed and cannot be bent at right angles.

B) Unbounded Transmission Media

While it is not within the scope of this chapter to discuss all possible media, it is worth noting that transmission of information within a local area network is not restricted to bounded media such as wire, cable, or optical fibers. The techniques used for signaling over bounded media can be used, for the most part, over unbounded media.

Three major types of signals that are broadcast via un-bounded media are radio, microwave, and infrared. While there are currently few commercially available local area networks based on these technologies, they will undoubtedly become more prevalent over time.

In fact, one of the prototypical models for local area networks, the ALOHANET in Hawaii, relies almost entirely on radio transmission.

One transmission medium which has emerged from experimental status to wide-scale operational use is satellite transmission. Communication satellites orbit the earth in a geostatic position functioning as relay stations for earthbound microwave communications links. This is ideal for long-distance communications links since the high altitude of the satellites avoids various earth-level interferences. The satellite communications systems presently in use offer very high bandwidth at low incremental cost per bit per second. On the other hand, these systems require relatively expensive earth stations. The next generation of satellite equipment, which will become available in the late 1980's, promises to provide even cheaper communications with the use of higher satellite frequencies, thus making it possible to use smaller satellite antennas at earth stations. One of the other important advantages of satellite communication is that it is inherently broadcast in nature, i.e., one sender can communicate with a number of receivers. This has obvious advantages for certain types of applications.

Still another possibility is the use of radio frequency transmission. Long-wave radio systems have been used for many years in telecommunications. More recently, microwave transmission using the high end of the radio frequency range has entered use by common carriers, especially on long-haul communications links. This requires less-frequent reamplifications and no wires, permitting very high bandwidth communications. However, microwave transmission requires a line-of-sight path with relay towers usually spaced about 30 miles apart. Another use of radio communication is for mobile data communications on a broadcast basis. This is most appropriate for use in access to larger communications systems in which mobility is essential. Still another application for radio is the connection of large numbers of terminals within urban areas to communications systems. Here, the high bandwidth and broadcast nature of the transmission medium make it possible to connect a large number of users at relatively low cost.

1) <u>RADIO</u> Local network utilization of radio as a transmission medium revolves around packet radio technology. The advantages of multiple access and broadcast radio channels for information distribution and computer communications have been established, and several experimental digital radio networks are in operation. Packet-switched communications techniques are particularly important for computer communications in the ground mobile network environment.

Many technical problems exist such as allocation of radio channels. The choice of radio channels for any communication system is a complex task, requiring tradeoffs of many factors such as desired bandwidth, area coverage, spectrum availability, potential interference and noise sources, regulatory requirements, and frequencies where spectrum crowding is less severe and the availability of bandwidth is greater. Crowded radio bands are undesirable because of interference with other users. The packet radio signaling waveform must be designed to perform well with respect to both the natural environment and the induced environment arising from both intentional and unintentional interference. Such interference includes system self-interference arising from the multiple access/random access nature of the packet radio system. Spread spectrum techniques reduce the multipath-caused limitations on signaling rate, and provide rejection of interference and the ability to coexist with other signals in the RF band.

Two major types of spread spectrum signaling techniques are well-suited to packet radio applications: direct sequence pseudo-noise (PN) modulation, frequency hopped (FH) modulation,and hybrid combinations of the two. Bandwidth is expanded by these techniques, but additional advantages include:

- Signal to noise ratio is improved by a factor called the processing gain.

- Separation ability of the various multipath signal components, allowing recombination with reduced signal fading over time, improved signal to noise ratio, and frequency selective fading only over small portions of the band.

- Lower electromagnetic profile since the signal is spread over a wider bandwidth and its waveform is of a pseudo random nature.

- Strong ability of receiver to correctly receive one packet in presence of other interfering packets (capture capability).

- Users can coexist in same area of frequency band with reduced interference.

- Reduced overlap of signal with delayed components.

- Reduced multipath fading effects.

The operational characteristics of the radio frequency band have a major impact on the packet radio design. The lowest and highest frequencies that can be used for packet radio are determined primarily by consideration of bandwidth and propagation path loss.

Bandwidth limitations are imposed by the operational characteristics of the radio frequency band and network requirements. Cost-effective radio center frequency is much larger than 0.3. This lower bounds the range of acceptable RF center frequencies. In practice, a center frequency well in excess of this lower bound is desirable if the receiver would otherwise have too wide a multipath spread. For example, if a packet radio system is to deliver 2000-bit packets through a network with delays on the order of a tenth of a second, the data rate of the system must be in the range of a few hundred kilobits per second, implying bandwidths of a few hundred KHZ. From an implementation viewpoint, the RF center frequency should be at least a few MHz in the lower high-frequency band (HF) extending from 3 MHz to 30 MHz. Propagation in the HF band does allow long-distance communication but limits data rate due to multipath spreading. Line-of-sight propagation dominates in the VHF (30 MHz to 300 MHz), supporting data rates on the order of a hundred kilobits due to reduced multipath spreading.

By its nature, radio is suited to both broadcast and point-to-point connectivity. However, numerous technical problems still result from station mobility, propagation characteristics of the frequency band, natural and man-made interference, reflection, and noise.

Ground-based networks encounter the most difficult environment in terms of propagation and RF connectivity. Ground radio links are subject to severe variations in received signal strength due to local variations in terrain, man-made structures, and foliage.

Reflections give rise to multiple signal paths leading to distortion and fading as the differently-delayed signals interfere at a receiver. RF connectivity is difficult to predict and may abruptly change in unexpected ways as mobile terminals move about. The implementation of the packet radio system should have a self-organizing, automated network management capability which dynamically discovers RF connectivity as a function of time, providing area coverage with full connectivity.

The distance between points communicating by packet radio is highly dependent on the propagation characteristics of the employed frequency band. Propagation in the HF band can provide long-distance communication due to sky wave reflections from the earth's ionosphere. But the propagaton suffers from noticeable multipath spreading of the signal which limits the data rate. As the operating frequency rises to a practical upper limit of 10 GHz, absorptive losses due to the atmosphere and rain rapidly increase, reducing the resulting radio range. Since local networks have an intra-city scope of a few tens of kilometers, radio frequencies in the upper VHF (100

MHz to 300 MHz), UHF (300 MHz to 3 GHz), and lower SHF (3 GHz to 10 GHz) bands can be utilized to provide that range. However, closely spaced relays must be utilized to provide adequate area coverage at these frequencies.

Reflections caused by local variations in terrain, man-made structures, and foliage allow multiple signal paths to a destination, imposing distortion and fading as the differently- delayed signals interfere when received. Operational characteristics of the frequency band employed may cause multipath spreading, fading, and distortion, particularly at HF and lower VHF. The effect is that additional attenuation of the signal may be observed when the receiver is located at a ground point where the signal interference is destructive.

Radio is affected by both intentional (jamming) and nonintentional interference; requirements for resistance to interference strongly affect the details of modulation techniques selected and system complexity. Unintentional interference results from automobile ignitions, trains, radar, etc., and is often generated at relatively low signal strength levels. However, in urban areas the density of interference sources will be quite high, causing levels of 60 to 80 dB above the thermal noise level of the receiver. Packet radios might therefore experience bit errors in every packet received, requiring forward error correction to maintain system throughout.

Packet radio is a technology that extends the original packet switching concepts which evolved for networks of point-to-point communication lines to broadcast radio networks. Development has been greatly stimulated by the need to provide computer network access to mobile terminals and computer communications in the mobile environment. Because of its capability for dynamic allocation of the spectrum, packet radio offers a highly efficient way of using a multiple access channels, particularly with mobile subscribers and large numbers of users with "bursty" traffic. It can also provide a degree of flexibility in rapid deployment and reconfiguration not currently possible with most fixed plant installations.

Packet radio will be essential for military and other governmental needs as terminals and computer systems become pervasive throughout essentially all aspects of their operations. Initially, the needs for radio-based computer communications are expected to be prevalent in training on or near the battlefield and in crisis situations. The first operational systems will most likely be deployed for use in one of these areas where a higher relative cost of providing the advanced capability can be tolerated.

Within the civilian sector, there is also a strong need for terminal access to information in the mobile environment. But the cost of services to users will dictate when such capabilities should be publicly provided.

Based on the above discussion, it is clear that radio technology for local networks is still in its infancy and quite experimental. Present applications might involve point-to-point or store-and-forward transmissions between buildings or across a city in which transmitting/receiving equipment may be placed in direct line-of-sight to avoid urban effects.

Due to the experimental stage of packet radio technology, complete network costs would be quite high since extensive custom development would be required. Radio transmitters and receivers are readily available but total network capabilities are not presently available off-the-shelf.

2) HF OR FM RADIO SIGNALS Although VHF or FM radio signals which start and stop functions are becoming popular, problems have been experienced. The basic problem is obtaining frequencies due to the limited number available.

However, even if frequencies can be obtained, interference can be a problem, e.g., an airplane flying overhead can trigger a system. This problem can be avoided only through extensive fine-tuning.

Expandability also can be a problem. If 100 items are to be activated and deactivated at different times, 100 different frequencies and signals are needed.

Other problems include:

Signal distances are limited since longer distances increase the possibility of interferences from airplanes, CB radios, etc.

- High maintenance requirements due to the large number of transmitters and receivers involved.
- Low reliability.

In some cases, a combined system will be employed in which an FM radio signal and a carrier signal are carried on a power line. The carrier signal is transmitted through the low-voltage side of a transformer; the radio signal takes care of the rest of the transmission. A combined system is limited because of the large number of items normally transmitted on the low-voltage side. Scan time can be relatively fast but there is a limit to what can be scanned.

3) <u>MICROWAVE TRANSMISSION</u> Microwave transmission is a practical alternative for communication between facilities separated by considerable distances. The positive attributes of microwave transmission include a fast scan rate that is compatible with most requirements, excellent reliability (assuming knowledgeable maintenance personnel are available) and compatibility with future requirements and expansion.

The primary negative attribute of microwave transmission is high initial cost. Receivers and transmitters are needed in each building.

If microwave transmission is used, other types of data media should be specified for completing the trunk wiring system.

4) <u>INFRARED</u> In addition to fiber optics, information can be transmitted at infrared light frequencies through free-space or atmosphere. Light sources for infrared transmissions include both LEDs and two types of lasers. Gas lasers have the disadvantages of higher cost, greater bulk, and high voltages. By contrast, semiconductor laser diodes and LEDs provide low cost, small size, high efficiencies, a range of wavelengths and

long lifetimes. These devices also operate on low voltages and can be directly modulated but have a high degree of divergence, wider spectral widths and low duty cycles. A laser diode is better for longer ranges than an LED, due to its smaller divergence, larger peak powers and smaller spectral width which allows improved background noise filtering. The LED, however, is capable of the higher duty cycles necessary to achieve high data rates, longer lifetimes and lower costs.

Light is not only generated via the above sources but must also be detected at the receiving end. Direct detection is most practical for severe operating conditions in which atmospheric conditions perturb the phase and directionality of the wavefront. The most common detectors are semiconductor PIN photodiodes and avalanche photodiodes.

To protect against atmospheric effects and background noise, short pulse with high peak power modulation techniques are utilized. Pulse modulation has made a wide variety of coding formats possible, using both analog and digital inputs. Digital coding formats adapt well to pulse modulation, often with significant performance advantages over analog transmission, but do require more complex circuitry and larger bandwidths.

Pulse-rate modulation can partially overcome atmospheric scintillation effects. Pulse-position modulation (PPM) and differential pulse-position modulation (DPPM) offer the most promising benefits such as fewer atmospheric problems compared to other modulation techniques, low harmonic distortion (less than 2 percent out to 3 KHz), and low power requirements.

Most systems presently in operation use laser diodes and LEDs as sources with semiconductor detectors. Data rates, therefore, fall between 10 to 100 K bits/second with 10 error rates over a range of approximately 16 kilometers. Reliable data rates up to 1.5 M bits/second are possible over a 1.6 kilometer path. A number of optical communication link systems are presently marketed with data rates and distance closely tied.

Present free-space infrared communications technology is limited to point-to-point links (star, ring topologies). But recent advances make use of infrared properties that allow easy confinement of transmission within a desired area of reception, i.e., a room; in a true broadcast mode. A central infrared station installed in a room's ceiling is able to transmit and receive to and from specific points within the room, using diffused nondirectional radiation. Signals are reflected in all directions, filling the room and eliminating the need for line-of-sight transmissions.

The atmosphere imposes severe restrictions on the range and error rates for infrared communications due to absorption, scattering, turbulence, refractive index variations, and scintillation. This effectively limits optical communications through the atmosphere to ranges of less than 160 kilometers, and to less than 1.6 kilometers for the large bandwidth promised by light-wave communications. The average range is 1.6 kilometers and the practical maximum range is between 1.6 and 3.2 kilometers.

Infrared light transmissions are unaffected by electromagnetic interference and noise. But atmospheric scintillation modulates the intensity of the beams at 1 to 200 Hz rates and beam wander and spreading caused by refractive index changes along the transmission path will often cause sudden deep fading. Placing analog information on an FM subcarrier modulates the intensity of the laser beam, lessening the importance of nonlinear operation of system subcomponents, changes in the laser amplitude and atmospheric effects.

Free-space infrared communications applications have been primarily limited to point-to-point line of sight situations such as between closely located buildings. The technology is still in its infancy but broadening of the applicable areas is beginning to take place. Manufacturers have created terminal networks within a room in which an infrared source in the ceiling, for example, transmits/receives to/from terminals in the room using diffuse radiation. Advances are expected to continue and the medium may become viable for general network application in the future.

As might be expected, free-space infrared communications devices are quite expensive. Costs will drop as technological and manufacturing improvements are made but this will take time. No general network capabilities using this medium are available. Extensive custom development would be required to create and install such a network at the present time.

TRANSMISSION MEDIA APPLICATIONS IN BUILDING AUTOMATION SYSTEMS

In simple applications involving a single type of transaction and relatively small number of nodes, the selection of media will often become intuitively apparent from a tabulation of system requirements. Larger, more complex data communications systems may profitably rely on special models that assess the relative costs and performance characteristics of alternate multimedia applications. For each set of user requirements, there is a transmission technology which has the greatest potential for increasing value-added capabilities.

Distinctions between local network applications revolve around four basic issues: the type of information that is transmitted (data, voice), the type of equipment that requires communications (terminals, computers), advanced services that may currently be required by users or for which applications may evolve and any specialized environments.

An astute system designer will have roots in the parallel disciplines of communications and computers, and will design systems to conform to the unique characteristics of each individual application. The design of a data communications system involves a sequence of steps. Decisions at each step are intertwined with those taken earlier and later. Although design must often invoke formal, mathematical-based procedures, the planner's experience, intuition, and willingness to make a decision can go a long way toward final design. Also, the direction of competitive products lends weight to decision processes.

The key to meeting the evolving needs of users is to understand their communications applications and then apply appropriate technological solutions instead of arranging applications to conform to technological constraints.

CONCLUSIONS

The first step in the design of a communications system is to specify the design objectives:

- Performance levels

- Functions provided

- Applications supported

- Cost effectiveness

Then, transmission media selection criteria can be used to design systems to exploit the best features of each technology in terms of optimizing service, connectivity, compatibility, performance and cost. The determination of which communications medium is "best" for a particular application also depends on the environment and structural characteristics of a facility.

There is a strong probability that very large network installations will use a hybrid network or several different networks. A hybrid can provide maximum flexibility by using the most efficient medium for handling each kind of traffic--voice, video, and data.

In order for systems to have the capability to exercise the multimedia option without changeover problems, excessive cost and creation of a system without an identity, future design should be modularized to allow for alternative media and media-access techniques. As a result, only software modifications will be required to meet future user needs. Ideally, consistent hardware and software interfaces to networking services will be provided throughout a BAS product line and across a variety of physical media, allowing flexibility for growth and advances in communications technology. This will make it possible for application engineers to choose the medium or combination of media which will provide required capabilities now while retaining the option to use other media in the future.

APPENDIX 1 COMPARISON OF SIGNAL TRANSMISSION CABLES

CHARACTERISTIC	CABLED SINGLE WIRE	CABLED TWISTED WIRE	COAXIAL CABLE		TRI-LEAD	FLAT CABLE	FIBER OPTICS
			SOLID CORE	AIR-SPACED			
Impedance Tolerance	P	P	E	E	G	E	—
Attenuation	P	F	E	E	G	G	E
Crosstalk	P	G	E	E	G	F-G	E
Time Delay	P-F	P-F	G	E	G	G	P
Rise Time	P	F	G	E	G	G	E
Band Width	P	F	G	E	G	F	E
Mechanical Integrity	E	E	G	G	F	F	P
Flexibility	G	G	E	G	E	E	P
Cable Dimensions	P	P	G	E	G	G	E
Dimension Tolerance	F	F	G	G	G	E	E
Cable Cost	E	E	G	G	F	F	G
Installed Cost	G	F	F	F	F	E	P

from Belden Cable Company

APPENDIX 2   COMPARISON OF NETWORK TRANSMISSION MEDIA

	Twisted Pair	Baseband Coax Cable	Broadband Coax Cable	Fiber-Optic Cable
Bandwidth (H$_z$) Max	Low (100K to 1M)	Low-Moderate (10M to 50M)	High (300M to 400M)	Virtually Unlimited
Component Availability	Widely Available	Limited	Widely Available	Very Limited
Component Cost	Very Low	Medium Minus	Medium Plus	Very High
Interconnect Complexity	Low	Medium Minus	Medium Plus	High
Medium Cost	$0.20/m	$2.00/m	$1.00/m	$10.00/m
Multidrop Suitability	Fair (32-200)	Good (64-1024)	Excellent (200-24,000)	Poor (2-8) (Needs repeaters at each node)
Number of Nodes (Typ)	10s	10s to 100s	100s/Channel	2 (point-to-point)
Signal-to-Noise Ratio	Low-Moderate	Moderate-High	High	Very High
Technology State	Mature	Developing	Developing to Mature	Emerging
Transmission Distance (Max)	Low (10s of Metres)	Medium (2.5 km)	Medium-High (300 km)	High (100s of km)
Topology Versatility	High	High	High	Moderate-Low (bus and tree are difficult)
Ease of Installation	Moderate	High	High	Moderate
Cost of Installation	Low ($0.40-$2.25)	Moderate ($0.25-$4.00)	Moderate-High ($2.50-$4.00)	Very High ($2.00-$7.00)
Reliability	Very good	Excellent	Excellent	Excellent
Growth	Poor in Dedicated Plant	Poor-Fair	Excellent	Good in Point-to-Point

APPENDIX 3   OSI (OPEN SYSTEM INTERCONNECTION) REFERENCE MODEL

A seven-layered system developed by ISO/TC97/SC16 to allow communication between users without dependence on a particular manufacturers equipment. The seven layers are: physical, link, network, transport, session, presentation, and application.

APPLICATION
PRESENTATION
SESSION
TRANSPORT
NETWORK
LINK
PHYSICAL

APPENDIX 4   APPLICATIONS

TWISTED PAIR	• PROCESS CONTROL • NUMERICAL CONTROL • SOFTWARE DEVELOPMENT • VOICE & DATA CONCURRENT • WORD PROCESSING • MOST EFFECTIVE FOR SINGLE BUILDING LOW TRAFFIC LAN
BASEBAND COAX	• SAME AS TWISTED PAIR • REAL-TIME SIMULATION • LIMITED VOICE (CPU-CPU MULTIDROP BUS SYSTEM LACKS LARGE SCALE CAPABILITY IN PLANT DISTRIBUTION)
BROADBAND COAX	• SAME AS TWISTED PAIR • REAL-TIME SIMULATION • VOICE • VIDEO CONFERENCING • (LARGE CAPACITY SUPPORTED BY HIGH MULTIDROP CAPABILITY IDEAL FOR DATA DIST. NETWORKING)
FIBER OPTICS	• CPU-CPU • CPU-HS PERIPHERALS • VIDEO CONFERENCING • SENSOR MULTINETTING • REMOTE DEVICES • TELCO CIRCUITS (LONG DISTANCE TRUNK) • HIGH NOISE IMMUNITY • GOOD FOR POINT-TO-POINT LINKS BETWEEN BUILDINGS

APPENDIX 5   TRANSMISSION MEDIA

	CAPACITY	INTRA BUILDING	INTER BUILDING	INTRA CITY
TWISTED PAIR	10 MBPS	X	X	
COAX CABLE (BASEBAND)	50 MBPS	X		
CATV CABLE (BROADBAND)	20 MBPS	X	X	X
MICROWAVE	1.5 MBPS		X	
FIBRE OPTICS	250 MBPS	X	X	
LASER	1.5 MBPS		X	
INFRARED	250 KBPS		X	
FM RADIO	19.2 KBPS	X	X	X

FROM MANAGEMENT DEVELOPMENT FOUNDATION, INC.
DATA COMMUNICATIONS WORKSHOP

Chapter 13

Premises Distribution Systems

Gerald F. Culbert, President
Information Transport Systems, Inc.

Why are premises distribution systems important?

In the past, you didn't have to worry much about communications distribution systems. When you ordered your telephones or PBX system, for instance, the distribution system to support them was usually "part of the package." No doubt you paid for it in some way, but the costs were as invisible as the decisions involved in installing it. Often, because these costs were distributed among many products and customers, there was no way of knowing the real costs for the wiring and labor.

As a result of FCC meetings, however, this situation has changed. Much of the wiring you have taken for granted on your premises may have to be bought or replaced. Or, if you are planning a new building or campus, you now have to give some hard thought to choosing a distribution system which will be both functional and cost effective.

More than ever before, then, what to do about your wiring has become a complex issue, an issue that affects both your building and your business.

Before you can make any decisions about wiring, however, you have to know more about distribution systems and what they do for you. If you already occupy a building or campus, for instance, you have to know what wiring you already possess, and how much of it is reusable. And whether you are planning a new building or campus or occupying an existing one, you need to know about the functionality of various distribution systems. How serviceable is the system you have or are planning? Does it allow room for growth? Does it have the flexibility and the communications capacity you need to accommodate your demands for voice, data, energy, security and fire? Will it be able to meet your future needs? Or will it have to be changed or upgraded every time you switch to a new system? Can it be administered easily, or does it require trained service personnel to make the necessary circuit changes when equipment is moved from one office to another or one area to another? What sort of warranty accompanies it? How easy is it to obtain reliable service? Answering any of these questions requires some familiarity with the terms, components, and capabilities of distribution systems.

WHAT IS A DISTRIBUTION SYSTEM?

A distribution system is the cables, adapters, and other supporting equipment that connect telephones, data terminals,

Fig. 13-1 Distribution Subsystem

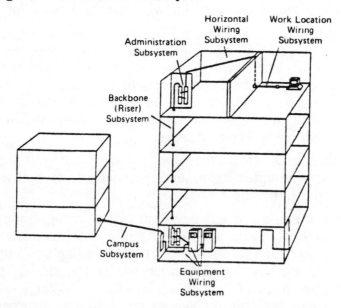

and energy, security, and fire management devices, allowing them to talk to one another. As such, a distribution system is an assemblage of various products, many times from multiple vendors.

But a distribution system is also a method of arranging these products within a building or on a campus in a logically coherent and economical fashion. After all, equipment alone does not make a system any more than building materials make a building. To meet this end, a distribution system can be organized as a set of subsystems tailored to specific needs.

There are six distribution subsystems:

- Campus
- Backbone (Riser)
- Horizontal Wiring
- Work Location Wiring (or Sensor)
- Equipment Wiring
- Administration

The campus subsystem (Figure 13-1) consists of components, such as cables or a microwave system, that support communications between buildings on a campus or in an office park where there are no right-of-way problems.

The backbone (riser) subsystem is the central or feeder group of cables in a building. In a multi-story building, these cables typically run from an equipment room to other floors where they are terminated in backbone (riser) closets. In a one-story building spread out over a large area, however, the backbone cables run horizontally. The campus and backbone cable is sometimes called "black cable" because of the color often used for its sheath.

The horizontal wiring subsystem is the wiring or cabling that runs from a backbone or satellite closet on each floor to an information outlet (such as a wall jack) in a user room. In some instances, this cabling can also be used to feed multiplexers or actual sensors in an energy or security management system. This cabling is sometimes called "gray cable," again, because of its typical sheath color.

The work location subsystem consists of the mounting cords and connectors that link station equipment, such as telephones, data terminals, workstations, or some type of sensor, to the information outlets in rooms and offices.

The equipment wiring subsystem consists of the cables and connectors that link voice and data switches, host computers, and other devices to shared equipment or cross connections in the equipment room.

The administration subsystem permits communications circuits to be routed and terminal equipment to be relocated without extensive rewiring. The three means of administering circuits are with cross connections, interconnections, and with information outlets, typically wall jacks in user rooms, that permit equipment to be disconnected simply by unplugging it.

WHAT'S IN A DISTRIBUTION SYSTEM?

Distribution subsystems are created by combining components from the following categories:

- Transmission Media
- Cross-Connect and Interconnect Hardware
- Connectors, Plugs, and Jacks
- Adapters
- Transmission Electronics
- Surge Protection Devices
- Support Hardware

Not every system will have all this equipment. Some, for instance, will have no need for transmission electronics. Others will not require protection devices. But every system can be composed from the types of equipment on this list.

Transmission Media

A few years ago, this section could have simply been called "wiring." Now, however, the term "wiring" is developed out of materials such as glass. Consequently, although terms like "distribution media" or "transmission media" are less familiar, they are more accurate.

Some transmission media can carry more traffic than others - - like roads with two, four, or eight lanes. The information carrying capacity of a medium is roughly proportional to what is called "bandwidth." Bandwidth refers to the range of frequencies (measured in hertz or cycles per second) that a medium can accommodate without significant interference or loss of signal strength. The greater the range of frequencies, the more information a medium can handle.

The most commonly used transmission medium is twisted pairs. A twisted pair consists of two copper wires, each in its own color-coded, insulated covering, twisted around one another to minimize electrical interference. These are usually combined with other pairs in an additional sheath of plastic or other material to form a single cable.

Twisted pairs are the kinds of wires that usually connect your telephone, data terminals, energy or security sensors. Most of your standard telephones and data terminals use one, two or four pairs (two, four or eight wires). But older telephone sets can use as many as 25 pairs.

Between the floors of buildings where there are many telephones, data devices or sensors, there are usually large cables of 100, 300 or more pairs.

There are also other kinds of copper conductor cables - not necessarily twisted pairs - like flat or ribbon cables, useful for short runs or under carpets where exposed cables are undesirable. Twisted pairs remain the most commonly used distribution medium because they are inexpensive to buy or lease, relatively easy to pull into tight wiring channels and conduits, and versatile in their applications. In fact, twisted pairs are suitable for most voice, data and sensor applications.

The most promising transmission medium for wide band applications is optical fiber (see Figure 13-2). Optical fiber is essentially a glass filament designed to carry lightwaves with minimal dispersion. It is sometimes called a "lightguide" and perhaps this term better than any other suggests how optical fibers work. An optical fiber guides a coded series of light pulses down a path so that none of the information-carrying light signals gets lost.

As a transmission medium, optical fiber has a number of advantages over both twisted pairs and coaxial cable:

- Greater Bandwidth
- Smaller Size
- Lighter Weight
- Greater Security
- Potentially Lower Cost

In addition, optical fiber is neither susceptible to, nor does it create electromagnetic interference.

Fig. 13-2 Optical Fiber Cable

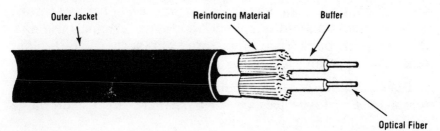

Optical fiber's other qualities make it particularly appropriate in complex electrical environments. Since glass is a nonconducting material, it is not effected by the electromagnetic fields created by nearly current carrying wires. There are no crosstalk, static, or other electromagnetic noise problems with optical fiber.

Similarly, optical fibers do not create interference themselves. Consequently, they are compatible with other signals. They can be placed virtually anywhere and the need for costly shields is eliminated.

Cross-Connect and Interconnect Hardware

If people never moved from one office to another, there would be no need for flexibility in distribution design. Cable could be run directly to a device and never have to be changed. In most buildings, however, change is the rule rather than the exception, and a distribution system should be installed in ways that will allow for easy rearrangement.

An important feature in such a distribution scheme is the use and location of administration points. These points are a kind of switching yard where a circuit (a communication path between devices) can be routed or rerouted to reach a desired location. For example, a circuit which had formerly taken one path can be physically altered here so as to reach a new destination: when your office and phone are moved, the circuit can be administered so that you still keep your old telephone number.

This circuit administration for twisted pairs can be provided by either cross connections or interconnections (see Figure 13-3). Cross connects are panels, blocks, or fields where wiring from various sources is brought together so that connections can be made between them or "across" them. Cross connections for twisted pairs use either jumper wires or patch cords.

A jumper wire is a short length of wire that connects two terminal locations on a cross-connect field. Normally, two or more jumper wires are required per circuit. Both the jumper

Fig. 13-3 Interconnect Administration

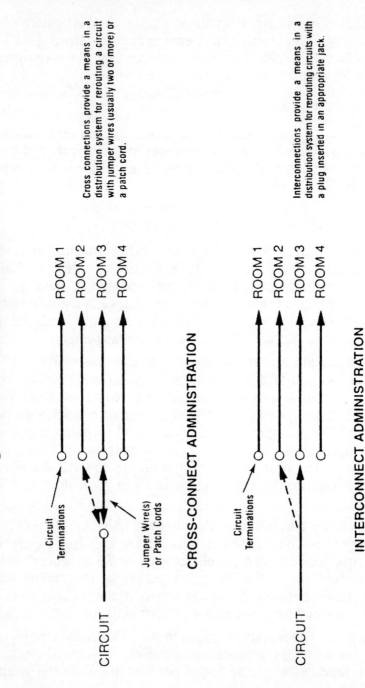

CIRCUIT

Circuit
Terminations

Jumper Wire(s)
or Patch Cords

ROOM 1
ROOM 2
ROOM 3
ROOM 4

Cross connections provide a means in a distribution system for rerouting a circuit with jumper wires (usually two or more) or a patch cord.

CROSS-CONNECT ADMINISTRATION

CIRCUIT

Circuit
Terminations

ROOM 1
ROOM 2
ROOM 3
ROOM 4

Interconnections provide a means in a distribution system for rerouting circuits with a plug inserted in an appropriate jack.

INTERCONNECT ADMINISTRATION

wires and the cable pairs they link together are secured to terminals in the cross-connect field by means of a push-down tool.

A patch cord contains several wires (usually enough for one circuit) terminated at each end in a pluggable connector. Cross-connect hardware using patch cords is referred to as a patch panel (see Figure 13-4).

Like cross connections, interconnections for twisted pairs allow circuit rearrangements; but, unlike cross connections, they don't use jumper wires or patch cords. Instead, the two ends of the circuit are directly connected by devices, such as jacks and plugs, usually mounted on an "interconnect panel," which provides an orderly means of rearranging and identifying connections.

Fig. 13-4 Panel For Wire Cross Connection

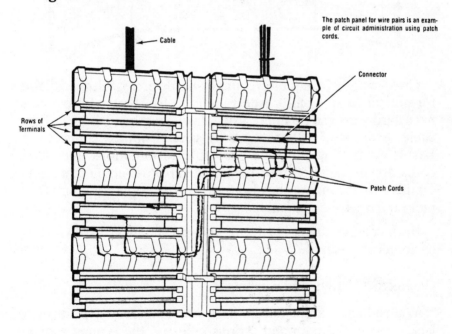

The patch panel for wire pairs is an example of circuit administration using patch cords.

Cable

Connector

Rows of Terminals

Patch Cords

Fig. 13-5 Panel For Optical Fiber Interconnection

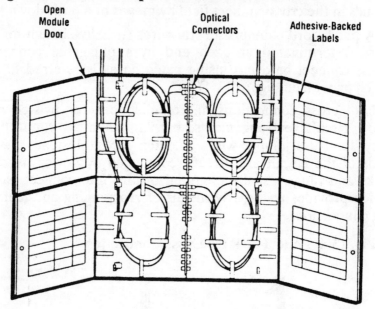

Cross-connect and interconnect hardware is also available for optical fiber circuits (see Figure 13-5). Optical cross-connect hardware uses patch cords (short lengths of optical fiber cable terminated at both ends by special optical connectors). Panels containing optical fiber cable terminations and patch cords are usually referred to as optical cross-connector units. Optical interconnect hardware provides optical connectors mounted on a special panel.

Both cross-connect hardware and interconnect hardware are customer serviceable.

Connectors, Plugs and Jacks

Where there are cables, there are connectors, plugs, and jacks; and in the communications industry, they come in many shapes and sizes. Although there are no hard and fast rules differentiating connectors and plugs, you will probably be safe calling devices which attach to equipment or cross-connect fields by the term "connectors," and the devices which are inserted into information outlets (wall jacks) by the term "plugs."

To give you a better sense of these devices, let's look at the connectors, plugs, and jacks you might find at various locations in your distribution system. At the top of the cross-connect field in the equipment room, there might be a number of 50-pin connectors for connecting the 25 pair cables from the communications equipment.

The 8-pin modular information outlets in your office terminate the cables from a satellite wiring closet. It's into one of these outlets that you insert the modular plug on the cord from your telephone.

If you have a data terminal on your desk, the cord plugged into one of the modular jacks (information outlets) probably goes to a data interface device (such as a modem), which in turn is connected to a D-Connector associated with your terminal. This is a 25-pin connector whose pins or leads are assigned according to conventions established by the Electronics Industries Association (EIA).

Fig. 13-6 A 50-Pin To Six 8-Pin Adapter

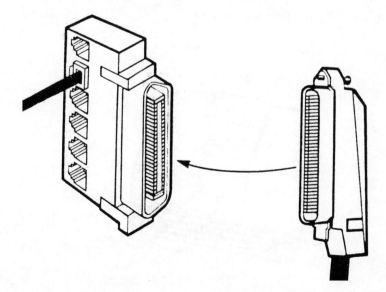

Adapters

There are several different types of adapters: those that enable incompatible plugs or connectors to mate with one another, those that allow a single large cable to be attached to a number of small cables, and those that protect the terminal equipment from undesired electrical conditions (see Figure 13-6).

Sometimes the obstacle to connecting devices is a purely physical one: the plugs or connectors are a different shape or they belong to the same gender. In this case, an adapter simply enables the physical connection to take place. At other times, the obstacle is more subtle: the connectors may be compatible but the signals on the individual pins or conductors may not be. One device may be sending control information over pin number five, for instance, while the other device expects that information over pin number six. In this case, an adapter may be needed to reroute the signals to the pin expected by the receiving device. Null modem adapters and crossover cables are examples of devices that perform rerouting function.

Fig. 13-7 Protective Adapter

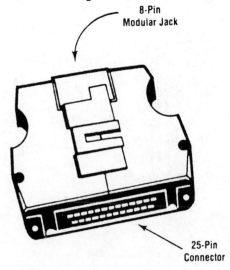

8-Pin
Modular Jack

25-Pin
Connector

Another type of adapter acts as an intermediary between large and small cables. Let's say you have a 25-pair cable running from a wiring closet to a large room. The easiest solution to fanning out all fifty wires to a series of 8-wire outlets is to use an adapter that attaches to the 50-pin connector at the end of the cable and provides female connections for six 8-pin modular plugs. Protective adapters are designed to prevent unwanted voltages in connecting cords from causing harm to terminal equipment (see Figure 13-7).

Transmission Electronics

Transmission electronics is a term that covers a broad spectrum of devices that repeat, reshape, or reformat signals from a transmitting source so that they can be more accurately or more economically carried to their destination.

One class of these devices is the repeater that amplifies and, in some cases, reshapes signals that have become weakened or distorted in the course of transmission. Repeaters that reshape digital data signals are called regenerative repeaters. Voice-frequency repeaters are used for voice and voiceband data transmissions.

Modems (modulator-demodulator) are also considered to be transmission electronics. Modems translate the digital signals produced by most computers into the analog or audio form expected by the telephone network, and vice versa. They thus allow computers to use the telephone lines either inside or outside a building to carry data from one location to another.

Still another class of transmission electronics is that of multiplexers, or "muxes" as they are sometimes called. These devices take signals from a number of different sources and send them over a common broad-band channel. They reduce the number of cables running between devices, and make more efficient use of transmission hardware.

Fiber multiplexers that perform the same concentrating function as traditional multiplexers, but translate electronic signals into optical pulses and use optical fiber as their medium are not available.

Surge Protection Devices

In varying degrees, almost all electronic equipment is susceptible to damage from surges produced by lightning on communiction cables and power lines. Also, where twisted pair cables share the same poles with power lines, a downed power line can be in contact with wire pairs and send high voltages through lines and equipment. Consequently, in most distribution systems you will find devices such as voltage limiters and current interrupters that protect electronic equipment against lightning and power surges on twisted pairs entering a building. The telephone company provides some protection for their twisted pairs entering a building, but you have to supply whatever additional protection is needed for these and for interbuilding cables in a campus environment.

DESIGNING AND IMPLEMENTING A DISTRIBUTION SYSTEM

Now that you have become acquainted with distribution subsystems and their basic components, we can discuss some of the steps involved in the design and implementation. These fall into four phases:

- Architectural Design
- Distribution Design
- Provisioning
- Administration and Maintenance

In such a brief introduction, our aim is to start you thinking about the general concerns that determine the design and implementation of a distribution system, not to lead you through any of the steps in detail.

Architectural Design Phase

The design and implementation of a distribution system is a complex operation. It should really begin at the same time that plans are being developed for a new building, addition, or major renovation. Since the initial phase of planning for a distribution system in any premise involves the actual physical design of the premise, we call it the architectural phase.

Assessing Your Needs

The first question to answer in planning any distribution system is, "What do I need?" If you are the head of a business, this means asking yourself what kinds of communications services you require to maintain and increase your viability as a company. If you are the owner of a building, this means projecting the needs of your probable tenants. In both cases, it means thinking about the future as well as the present; for a distribution system ought to have a much longer life span than the present communication systems it supports. A well-planned distribution system should accommodate not only the simple expansion that occurs when more of a building is occupied or more people are introduced into the same work area, but also the sort of upgrading that occurs as new technologies are developed.

To a large extent your needs depend first upon the type of business or premises you have. A hospital has very different communication needs than an office building, a retail establishment, a university, or a residence. So even if the amount of floor space were similar, the distribution system required in each case would be very different. Another factor will, of course, be the current and projected size of your premises. There are distribution options, such as the use of several administration points, that are cost effective for a large system but that may not be for a small one.

Finally, your needs will be affected by the status of your premises. Is it new or old, a planned construction or an already existing building? In a new building or premises, you are much freer to design a distribution system that precisely suits you. In an existing building, on the other hand, there are constraints that arise from the physical design of the building, the existence of other tenants, the desire to reuse wiring already in place - all of which will affect your distribution design. In this case, then, assessing your needs will be intimately connected to a careful survey of what you already have.

Playing It Safe

A prime consideration in any distribution plan has to be the protection of personnel and equipment from electrical shock and fire hazards. Before you begin designing your system, you must consult the relevant state and local building codes. These codes specify such things as material and construction standards, fire protection safeguards, and grounding and electrical protection requirements.

These codes are normally based on the National Electrical Code (NEC). But many states and localities require even tougher standards and controls. So your state and local building inspectors are the final arbiters on this subject.

Once you have assessed your needs and consulted state and local building codes, you must integrate that knowledge into the design of your premises.

Planning For The Service Entrance

The planning of a distribution system should begin at the same time as the architectural plans for the site. The first question that must be answered is where the service entrance will be; that is, where the telephone company's cables will cross the property line and enter the building. Are there, for instance, obstructions such as paved areas that would be encountered in one proposed site that could be avoided in another?

Other similar questions will need to be addressed. If you have several buildings, where is the most logical point of entry? And how are the cables going to get from one building to another? Buried in a trench or through a special conduit pipe? Above ground via poles? Each method has its advantages and disadvantages, and the method you choose will probably depend upon such factors as weather and soil conditions, right-of-way, aesthetic appearance, or just plain dollars and cents.

Once cables from the telephone company's central office have entered your premises and have been routed through protective devices, they are terminated on special jacks called the Network Interface. These jacks are the boundary between telephone company lines and those belonging to your premises distribution system. Anything that happens on the telephone company's side of the Network Interface is their responsibility. Anything that happens on your side of the Network Interface is your responsibility.

Allowing Space For The Equipment Room

As its name suggests, the equipment room is the place set aside for the communications equipment shared by many users. This includes cross-connect fields and telephone switching apparatus such as a PBX or data switch. In some cases, if telephone company policy permits, it may include the Network Interface, where telephone company cabling and your distribution system meet. The room may also be shared with the computer equipment. In a multi-tenant building, there is likely to be an equipment room for each tenant.

Since it is the location of cross-connect fields, the equipment room is also a place where circuit administration can take place. There are several questions to be asked about this room. Is it large enough to accommodate all the apparatus and cross-connect panels? Is it properly lighted? Does it have the climate control required by equipment manufacturers? Are there sufficient power outlets? Is the room properly

Fig. 13-8 Typical Riser/Satellite Closet Arrangements

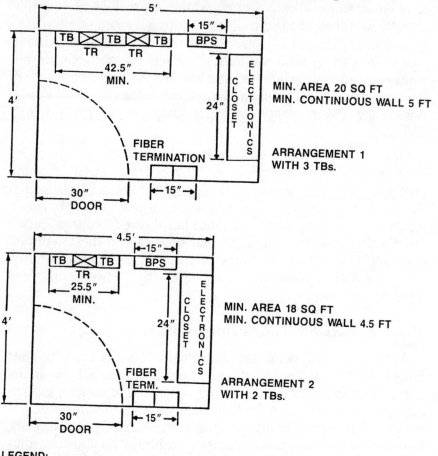

MIN. AREA 20 SQ FT
MIN. CONTINUOUS WALL 5 FT

ARRANGEMENT 1
WITH 3 TBs.

MIN. AREA 18 SQ FT
MIN. CONTINUOUS WALL 4.5 FT

ARRANGEMENT 2
WITH 2 TBs.

LEGEND:
TB - TERMINAL BLOCK
TR - WIRE TROUGH
BPS - BULK POWER SUPPLY

protected from flood or fire? Or, if you are planning an addition to an existing building, does your present equipment room need to be enlarged or altered to make it usable for the future? These are all questions that can be easily answered or resolved if you plan your distribution system early, in conjunction with your architectural designs. Answering these questions later, however, when a building or addition is already complete, can be costly.

Making Space For Wiring Closets

At the same time that you plan for an equipment room, you should also provide space for wiring closets (or cabinets in smaller systems). Wiring closets are of two kinds:

- Backbone (riser) closets

- Satellite closets

Backbone (riser) closets are located on each floor of a multistory building or along the central group of feeder cables in a single-story premises. Here the backbone cables are terminated and distributed to individual rooms or to satellite closets (see Figure 13-8). Satellite closets are the closets (or wall-mounted cabinets) that receive cables from a backbone closet and distribute them to individual rooms. Occasionally, with very small systems, cable can be run directly from the equipment room to user rooms, thus avoiding the necessity for these closets.

Let's assume you have a 4-story building, and you need 400 pairs for each floor. This could mean that you have a 1600-pair cable running vertically to all floors with 400 pairs brought out on each floor, or that you have four separate 400-pair cables, each going to a different floor. Each of the 400 pairs will be fanned out, probably in a closet, onto a cross-connect field. In the same location, there will be similar terminations for cable pairs from information outlets in user rooms. Several planning issues surface at this point:

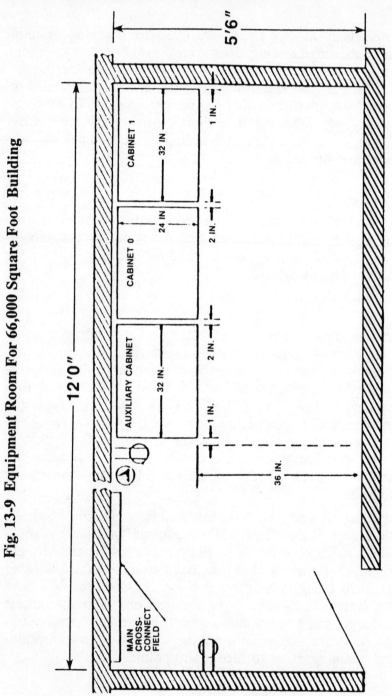

Fig. 13-9 Equipment Room For 66,000 Square Foot Building

- The wall space in the closet must be large enough to house the necessary cross-connect units.

- The closet should be centrally located in order to serve the entire area (some devices have restricted signalling distances.)

- If the backbone closet isn't sufficient, satellite closets will have to be planned.

- Future growth may require additional cable or equipment or both (and thus planning for their location).

- Closets should be secure, and yet easily accessed.

- The equipment served by this closet may need adjunct power supplies and thus additional adequate outlets.

- Multiplexers, if required, will need space in the closet (see Figure 13-9).

Choosing Distribution Methods: Backbone (Riser)

In order for cables to pass from floor to floor in a multi-story building, there has to be an opening in the floor between the backbone closets on each level. These must be planned for in advance so that the structural floor is not weakened by having to drill holes at a later date.

There are three commonly used methods for providing this opening: slots, sleeves, and conduits.

Slots are not much more than rectangular openings in each floor which enable the cables to pass through.

Although inexpensive, they must be "fire stopped," that is, plugged with fire retardant materials, to prevent them from acting like a chimney in a fire, spreading smoke and flames from floor to floor. In some buildings, however, it is impossible to have vertically-aligned closets. The backbone cables, then, must take a more circuitous route up the building. Whatever method you choose, you will have to face the same basic questions. Do you have enough space to accommodate the cables you need now and in the future? Will your system meet applicable safety requirements? Will it be easy to maintain?

Choosing Distribution Methods: Horizontal

Choosing a distribution method for the backbone or riser system must be done while a building is being planned, for the location of the riser shaft is part of the basic architectural design. The choice of horizontal distribution methods is less clear-cut. Some, such as the cellular floor method, must be planned as part of the building's very structure. Others, such as the raised floor method, can be planned and even installed after the building is complete. Planning in advance, during the architectural phase, however, is the best way to be assured that you have a full range of economical choices.

There are basically two ways of distributing cables on a single level: in the ceiling and in the floor; though for each alternative there are several different means which can be used.

In choosing a method, you will have to weigh the importance of safety, flexibility of rearrangement, initial cost, cost of later additions or alterations, and general aesthetics. And here again, your priorities will probably depend upon the kind of premises you have and the amount of service you need. The services you require may vary, in fact, from floor to floor or from area to area in a single building. Consequently, you may find one distribution method appropriate on one floor, and another distribution method preferable on another.

Ceiling distribution systems in general have their own special concerns. There are space requirements to be met and building codes to be observed. Ceilings often contain electrical power wiring, and there are strict regulations which govern the coexistence of electrical and communications wiring. In addition, if the ceiling area serves as an air return path for heating and air conditioning, cables must either be enclosed in an approved conduit or channel or have special low-flame, low-smoke properties. Finally, there are problems of physical support, since the more cables you have, the greater their combined weight.

There are at least four basic methods of distributing cables in the ceiling.

- Zone
- Conduit
- Raceway
- Poke-through

The zone method divides the usable floor space into zones or areas and runs all the cables (sometimes in rigid conduit) from the nearby backbone or satellite closet to the center of the zone. From there the cables can be fanned out or distributed to walls or utility columns. Its principal advantage is the flexibility for rearrangement that it offers.

The strict conduit method runs each group of cables in a continous conduit from the backbone or satellite closet to the desired outlet. As a method it's very direct and offers complete protection for your cables. But because it allows only one route for each set of cables, it is difficult and expensive to rearrange (see Figure 13-10).

The raceway method is like a series of interconnected suspension bridges that carry cables instead of cars. Larger raceways (headers) bring the cables into the center of the area; feeder raceways distribute them to user locations. The raceway method is similar to the zone method. But it is used in larger buildings or where the distribution system is complex enough to demand the extra support.

As its name suggests, "poke-through" is not the most sophisticated method. It consists of running cables in the ceiling of the level beneath your own and "poking through" (i.e., drilling through) to your floor. It is an inexpensive alternative for the short term, but if done often enough can weaken the structural floor and create a fire hazard.

The other major alternative to putting cables in the ceiling is to install them in the floor. The three methods most often used here are:

Fig. 13-10 Conduit Method

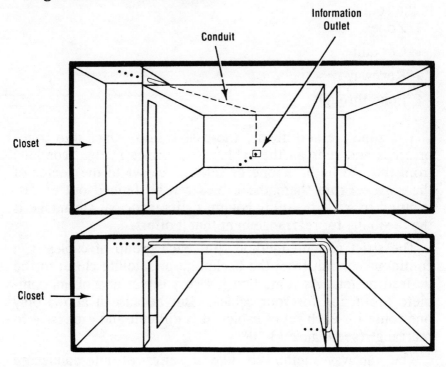

- Underfloor duct
- Cellular floor
- Raised floor

An underfloor duct system consists of a series of covered metal channels very much like the heating ducts in your house. And as in your heating duct system, there are straight ducts, curved elbows, and junction boxes where ducts intersect at right angles. Those junction boxes also serve as pulling and repairing points and are usually accessible through your tiled or carpeted floor.

Depending upon the complexity of the system, underfloor duct systems can be on one level (like your typical city street system) or on two (like a subway or freeway system). In either case, the system is safe and relatively flexible; but it does not require planning since it is usually installed while a building is under construction.

The underfloor duct system is purely a distribution method. It lies beneath, and is not part of, the floor. A cellular system, on the other hand, is part distribution method and part structural floor (see Figure 13-11). It consists of a series of flattened hills and troughs, giving it the effect of corrugated metal roofing. Through these hills and troughs, however, cables can be passed. Since, in most cases, the hills and troughs are less than a foot apart, this means that a cable can be run down the length of a room for easy access to any device. The cellular system is usually used in conjuction with a duct system of some sort to make the perpendicular connection between cells.

The most flexible of all floor systems is the raised floor system that you usually find in computer rooms. For the whole floor stands, in effect, on pedestals, and any section can be removed for access to the wire beneath. No system is perfect, however, and although raised floor systems make it easy to install, repair, and reroute cables, they are expensive, and sometimes create a "sounding board" as you walk on them.

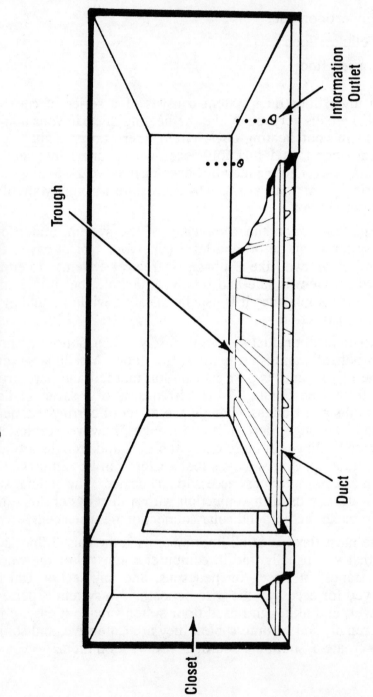

Fig. 13-11 Cellular Floor Method

Distribution Design Phase

In the architectural design phase, you have been preparing a building or premises for a distribution system by making sure that it contains all the appropriate facilities. You have made it easy to install a distribution system - whatever its kind.

In the distribution design phase, on the other hand, you are no longer designing a building to accommodate a distribution system, but designing a distribution system to suit the specific applications in your building. This is the phase when you will actually decide the type and quantity of cabling and cross-connect hardware you will need, where they will be located, and from whom you will purchase them.

If you have not already done so, you will want to carefully review your needs. If you are moving an existing system from an old to a new building, for instance, you would want a complete survey of all the telephone, telephone switching equipment, data terminals, host computers, data switching, and multiplexing equipment included or intended to be included in your new building. You would need to know the wiring and power requirements of all these components including the type and gender of their connectors.

If you were designing a distribution system for an existing building, you would also want to determine the amount of cable you already have.

Or if you were designing a distribution system for a building whose tenants were still unknown, you would want to establish the typical requirements of users in that type of building.

These surveys would form the basis for choosing the appropriate media, the number of cables in each location, the number and size of the cross-connect hardware, the number and type of connectors, plugs, jacks, adapters, multiplexers - in effect, all the equipment needed for your system. Since it is difficult to anticipate the most efficient and economical way to design a system, you would probably try several alternative ways, comparing the final results in each case.

After the distribution system has been designed, and, presumably, a vendor chosen to supply the products, you have reached what we call the provisioning phase of distribution design and implementation, the time when your equipment should be arriving on the dock. Certain questions will automatically arise here. Will the vendor you have chosen be able to provide all the products you need? Or will you have to deal with two or three different vendors to get the different kinds of equipment your system requires?

You may need to confront "multiple vendor" issues. If you deal with several companies, will they be able to meet your schedule, or will you find yourself with all the cables from one vendor but nothing to hook them up to because the cross-connect vendor did not meet the schedule?

You may also be concerned about product support. When your equipment arrives on your doorstep, will it be accompanied by installation, service, or maintenance manuals to help you use the product? There are few things more frustrating than having first-class equipment that you can't use because you can't understand how. Is customer training part of the package with the vendor you've selected? Or is that kind of service extra or nonexistent?

You will surely also have to look at installation issues. Who will install your distribution system? Will it be the same company who supplied the product? Or will you have to go to an outside firm for installation? Will you again find yourself in the middle of a finger pointing contest, as one accuses the other if the system doesn't work?

Finally, you will probably want to look into the question of testing and certification. Who will make sure that your cabling works once it has been installed? And how will that be determined? Will the company that manufactured the cable guarantee that it will serve you? Or will the company that installed it?

Administration And Maintenance Phase

After the installation of a distribution system, you enter the final phase of the design and implementation cycle, the administration and maintenance phase. This phase includes setting up procedures for the circuit administration required when moving equipment from one location to another and the troubleshooting that is occasionally necessary when a cable is inadvertently cut or some other problem develops.

Part of the success of any distribution system is the result of proper labeling; that is, of being able to tell where a wire comes from and where it is going. At any given time, the system administrator has to be able to tell which circuits are in use, which are unused, and where both can be found. Does the distribution system you are using have a labeling plan? Is it easy to understand and use? Will it enable you to make rearrangements quickly and easily?

And how about serious problems? Do you have a warranty from your system vendor? What does it cover? How long will it last?

Do you get fast, reliable service when you have a problem? Can you obtain on-going maintenance from the vendor you have chosen, Or will you do everything yourself once the warranty has expired?

Chapter 14

Electrical Wiring Management For The Intelligent Building

Alan B. Abramson, P.E.
Syska & Hennessy

In the past, communications requirements for voice, data and control systems were met through the use of dedicated cables individually wired between their respective points of interconnection. As communications requirements have grown, designers have typically addressed the expanding cabling needs of buildings via the provision of flexible, typically expensive, raceway systems. The rapidly expanding need for new and varying point to point communications links has led to the situation where buildings contain a maze of unidentified, often abandoned, cable. This occurs because new cable is added and old cable is left to remain because of the high cost of cable removal. Therefore, the apparently large and flexible raceway system may become filled with such cable and, in fact, size and flexibility has helped to exacerbate the same problem which it was intended to solve.

A new approach to this problem has been developed, which is best called Cable Management, that addresses the needs of both a facility and a tenant in a manner that significantly reduces the life cycle cost of the provision of signal level communication services. The key to the concept of cable management as presented in this chapter is the understanding of the need to plan requirements, recognize commonality, design flexibly, and manage communications for the life of the facility. This new approach has resulted from an examination of future directions in communication technology as they relate to the much slower changes that occur in the building process.

DEFINITIONS

The term "Cable Management" as used in this chapter relates to a process and not to a physically defined system. A detailed presentation of this process appears later in the chapter.

Further definition must also be given to the item being managed. For the purpose of this chapter, the term "Cable Management" will be used to identify all types of signal level transmission media; this includes copper media of all sorts, such as coaxial cable or twisted pairs, fiber optic cables, and other links such as microwave or satellite channels. It is, of course, implicit in such a discussion that other distribution equipment related to the transmission medium such as interconnection equipment be considered in the overall management process.

It is also important to note that one generically accepted definition of the term "Cable Management" relates solely to the provision of a flexible raceway system. Acceptance of this narrow definition will inevitably lead to a facility where cable will be unmanaged to the point where the raceways become full as described earlier. At that point in time, gaining control of the situation will be extremely difficult because the magnitude of the unmanaged cables will be significant due to the size and flexibility originally provided. The provision of a flexible raceway system as the sole solution to cable management is a poor approach and, in fact, represents an approach which may be termed "cable management avoidance." "Cable management avoidance" results from a lack of consideration for the transmission media and an intent on the part of the designer to postpone its consideration to the point in time where responsibility for cable becomes someone else's problem.

SUB-SYSTEM COMMONALITY

Cable management is essentially a communications problem. The basic intelligent building systems for which communication is required have been described in previous chapters. The commonality among these systems relates to the transmission of the following signal types:

- Voice

- Data

- Images

- Video

- Other discrete signals (sensors, controls, etc.)

Ideally, the degree of communications commonality among these systems encourages the consideration of integration and uniformity with regard to planning, design and continuous management.

Prior to presentation of this formal process, three critical points of caution that are common to all intelligent building approaches must be expressed. They are as follows:

1. Integration of systems should never be attempted merely for integration's sake. One must always look to achieve a functional synergy when integrating. Examples of the functional synergy that should be sought in an integrated cable management approach include items such as reduced first cost, reduced life cycle cost, increased flexibility or expandability, or increased ease of maintenance. Practicality must be stressed, and solutions whose only benefits are aesthetic do not belong in the design and building process.

2. Cable management cannot be successful if related only to signal level transmission media. A consistent approach must be taken to the distribution of electrical power and cooling media as well. In addition, equipment related to the intercon-

nection of the network or cables must also be considered in the context of current and future space allocation.

3. There are inherent roadblocks in the design and building process and in the respective tenants' organization to integration. For example, in a facility being designed for a developer/landlord, the HVAC control systems would likely be part of the base building design. Typically, installation of cables for tenant communication requirements would be part of the tenant space design process. Therefore, integration of those systems for cable management purposes would likely be impossible. A further example exists in the organization where responsibility for different intelligent building systems is separated among different departmental managers. While it is becoming less common, there are still many organizations who separate the managerial responsibilities for voice communications, data communications, and EDP facility management. These organizational barriers must be addressed prior to the consideration of an integrated cable management approach.

THE CABLE MANAGEMENT PROCESS

The cable management process has three distinct phases. These phases are as follows:

- Strategic Planning
- Distribution System Design
- Data Base Management

Strategic Planning

As previously noted, the rapid expansion of cabling requirements within an organization has typically led to a crisis management atmosphere with regard to the provision of these services. Therefore, the typical telecommunications manager, regardless of his or her career path in the organization, has likely never had the opportunity to plan for growth. Furthermore, in the multi-vendor environment created by the recent regulatory and anti-trust rulings, it is difficult to find a single source of responsibility for the coordinated planning of transmission media requirements. It is, therefore, imperative that prior to the consideration of any facility engineering solutions, the transmission requirements for the present and future be studied and presented in a strategic plan for the user.

The strategic plan must address the use of communications as a tool to accomplish the mission of the user in the short and long terms. To accomplish this process, interviews must be held with those responsible for the direction of the respective organization from a business point of view; those in the responsible administrative and facilities operation areas; and those to whom responsibility has been entrusted for support of communications as a business tool.

Every business and/or organization has a different character with regard to its perception of communications as a tool. In fact, the strategic planning process often becomes a time for rethinking the business mission and for education of senior management regarding the opportunities that modern technology offers.

Corporate culture must be recognized and incorporated into the communications strategic plan. A strategic plan will ultimately result in a fairly definitive projection of the functions and/or systems directions related to voice, data, administrative services and facilities as they fit the organization's mission.

A critical part of the strategic plan will be the provision of a migration plan which relates tactical approaches to a schedule of implementation for the acceptance and integration of new technologies. It must be noted that this is not an engineering-oriented task. It is a management and business planning task which may be lengthy and painful, but is also absolutely necessary. Without such a plan, the risks associated with any given engineering solution may be significant.

Distribution System Design

A distribution system is the cables, adapters and other supporting equipment that connects telephones, data processing

Fig. 14-1 Distribution System Design

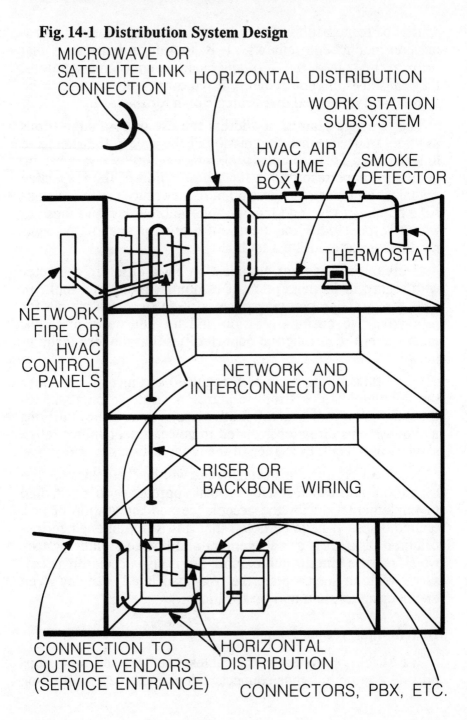

MICROWAVE OR SATELLITE LINK CONNECTION

HORIZONTAL DISTRIBUTION

WORK STATION SUBSYSTEM

HVAC AIR VOLUME BOX

SMOKE DETECTOR

THERMOSTAT

NETWORK, FIRE OR HVAC CONTROL PANELS

NETWORK AND INTERCONNECTION

RISER OR BACKBONE WIRING

CONNECTION TO OUTSIDE VENDORS (SERVICE ENTRANCE)

HORIZONTAL DISTRIBUTION

CONNECTORS, PBX, ETC.

devices, control panels and other communication devices, permitting them to "talk" to one another. (See Figure 14-1) The distribution system is represented by the assemblage of hardware in addition to the method of arranging this hardware within a building or campus in a coherent fashion. The intent of this interconnection is to future-proof the facility to permit the economical and flexible expansion of the networked devices without the need to significantly disrupt the facility physically.

The goal of a well-defined and designed distribution system is to permit integration of new requirements over time without disturbing floors, walls, and ceilings in the process. In addition, such a system should permit quick, inexpensive interconnection of all devices in a manner that suits the aesthetic requirements of the facility. The ingredients of a distribution system are the following sub-systems:

- Work Station Wiring

- Horizontal Distribution

- Riser or Backbone Wiring

- Interconnection and/or Network Equipment

The work station wiring connects the piece of equipment to a receptacle or jack on the wall or in the furniture. This cable is similar to a power line cord, is usually flexible and may be provided with the equipment.

The jack in the wall or furniture is connected through cabling that is permanently installed in raceways via a horizontal distribution sub-system. The form of this cable, in terms of the type and quantity of transmission media, will be determined based upon an engineering analysis related to the requirements stated in the strategic plan. Selection of the appropriate cable requirements is a critical part of a distribution system. Typical in modern distribution systems is the selection of multiple media which are distributed uniformly to all jacks in a facility. When jacks are placed at regular intervals and prewired as an in-place horizontal sub-system, the layout is typically called a uniform cabling plan. The essence of such a plan is the anticipation of short and long term media requirements and the slightly greater investment in providing these anticipated requirements prior to their actual need. The horizontal distribution system typically radiates from an interconnection point located in a communications room or closet.

The multiple closets in a facility are interconnected by a cable riser or backbone. The horizontal distribution system is interfaced to the backbone via interconnection hardware and/or network equipment. The interconnection hardware may be as simple as terminal blocks with patch cords or may be as complex as a rack of sophisticated local area network communications management equipment.

Media Selection

Media selection must be related to the migration path presented in the strategic plan. For example, if an organization intends to have a specific vendor's equipment for the long term, it should select a uniform cabling system supported by that vendor. If multiple vendors require support, then a hybrid approach may be appropriate. If high speed, high volume requirements can be anticipated, then a fiber optic approach may be appropriate.

It is also advisable to consider multiple uniform cables in the same facility if required. For example, in a health care facility, one type of uniform cable may be proper for the patient care areas while a different approach may work better for support areas. Typical media include twisted pairs (shielded and unshielded), coaxial cables, and fiber optic cables. Selection is based on an evaluation of the risk of not having cable available for a future vendor or tenant requirement versus the first cost of covering all contingencies. Therein lies the importance of the strategic plan and the criticality of an "artful" interpretation of the migration path.

User Sophistication

The beauty of a uniform cabling distribution system is its flexibility in dealing with multiple levels of sophistication and its adaptability to grow with a user over time. It may be as simple as a hard wired, point to point interconnection network within a limited location via "jumpered" connections within a closet. It may also permit point to point connections via cross connection on and off of a multiconductor backbone on different floors. When the user becomes more sophisticated, local area network equipment may be introduced where required by installing LAN equipment at the interconnection points. Sophisticated environmental control may be achieved by utilizing distributed direct digital control panels or fire alarm (smoke control) panels at these interconnection points.

Building Issues

Closets or communication rooms must be properly sized and/or designed for expandability to permit equipment requirements to grow.

With such a cabling system, flexibility of the raceways becomes a less important criteria, even though the requirement for its existence does not disappear. First cost for raceways will likely be reduced, even though initial cable costs are higher. Concurrent power and cooling media provisions must be made to make the plan work.

Data Base Management

Hopefully, it has become obvious by the previous discussions that the flexibility of a well-designed distribution system, combined with the multiple varieties of interconnected equipment, provides an extraordinarily large and changing mass of data with which management must keep current.

A Data Base management approach is absolutely necessary to avoid the problems noted in the beginning of this chapter.

While such an approach may be manual (by hand), a computer-based approach is recommended. Furthermore, there is a great advantage to utilizing computer-aided design technologies during the design period to create the data base for eventual management use.

CONCLUSION

Cable management is a multi-phased process that is evolutionary in its application. While good engineering is an integral part of the process, planning and continuous management are equally as critical to success and life cycle cost containment.

SECTION IV

Functional Integration

Chapter 15

Integrating Local Area Nets With Digital PBX's

Erik T. Ringkjob
InteCom, Inc.

The future of intelligent buildings rests primarily on the strategic applications of today's telecommunications technology. Users in multi-tenant environments are not only interested in the practicality of sharing a telephone system, but also in an economically feasible approach to accessing data through one common, high-speed network. From the most logical standpoint, these dual needs can be best satisfied by integrating local area networks (LAN's) with a single premise controller, rather than layering a building's communications backbone with individual systems for each of these desired features. The digital premise controller provides the major functions of a voice PBX coupled with variable speed data packet and circuit switching over the twisted-pair wires, centralized data base control and centralized maintenance.

To examine the advantages of integrating LAN's with the digital premise controller, it is important first to understand the definitions of each.

A true local area network was designed as a fixed transport medium to which many different data terminal devices can connect, allowing users to communicate in one homogeneous network. To be beneficial, a networking system must be capable of joining several other networks together and operate in an environment where resource sharing is easily accessible.

The traditional PBX was designed to establish voice connections only. The capability to switch voice, and both circuit-switched and packet-switched data, is not included in the offer-

Fig. 15-1 Shared Resource

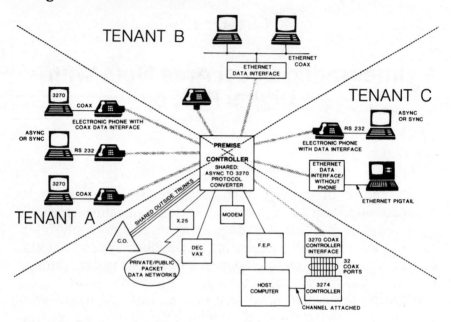

Illustration depicts tenants' access to simultaneous voice/data communications, circuit-switched asynchronous or synchronous data, 3270 coax elimination, virtual local area (ethernet) networking, protocol conversion and access to external wide area networks (packet-switched networks)--all supported by the integrated premise controller.

ings of the classic PBX. With calculated usage in mind based on the 6CCS (Six Centum Call Seconds) model, PBX's typically were designed with blocking architecture and driven by eight or 16-bit processors with limited memory. This created the capability to provide a predetermined volume of voice channels which could be utilized regardless of the number of stations attached. When that volume was met, additional telephone users were denied service by the PBX. Manufacturers designed the internal capacity around an allocated 600 seconds out of every hour, saying in essence that users could access their telephones for 10 minutes each hour. In voice-only situations, greater horsepower wasn't necessary. Users typically had ample access to telephone lines because, although there are many callers, none are on the phone for any great length of time; nor are there inordinate numbers of simultaneous conversations. Therefore, the blocking architecture of the classic PBX can be circumvented for voice applications. But, as we will see, the introduction of data to a PBX severely complicates the matter of continuous access.

INTEGRATING VOICE WITH OTHER TECHNOLOGIES

Hypothetically, a traditional PBX could be installed with 500 timeslots and 1,500 devices (telephones and trunks) could be connected to the system. Because of the blocking architecture of the PBX, only 250 internal conversations could be established simultaneously, thereby denying access to users once the timeslot capacity had been met.

Many PBX vendors have attempted to add data to their systems. These are not sophisticated data applications, but rather low-speed, typically 19.2 kbps or less asynchronous rates. A data device, added to a PBX, isn't used like a telephone. Traditionally, a data user, unlike a voice user who may use the telephone an average of 10 minutes per hour, needs the physical flow path from the terminal to the host computer for durations of several hours or more. When terminals with these long holding times are added to the blocking architecture of a PBX, timeslots are gobbled up in large chunks; a data terminal with a holding time of two hours per day demands the equivalent time slot allocation of twelve average telephones.

Expectations: The Integrated Premise Controller

PBX's typically support anywhere from 50 to 25,000 lines, with a technical limit to the number of data devices which can be added. The traditional PBX isn't the most viable tool with which to integrate LAN technology because of its non-blocking architecture, limited horsepower, processing capacity, random access memory (the ability of the PBX to store the data it needs to complete calls), narrow port bandwidth and overall system bandwidth. More often than not, traffic capacity would be exhausted before port capacity.

With this in mind, building owners would need to purchase three individual systems (voice PBX, data switch and LAN) and install separate wiring systems, manage separate data bases and create separate management disciplines to oversee those systems Today, with the proper "info-structure," or framework of information molded into one intelligent interface premise controller and LAN system, a building can offer tenants access to voice, data and local area networking from a single source. Initial benefits of such an integrated premise controller, really a current generation, high-powered PBX, are having one cabling plan versus three, twisted-pair wiring versus coaxial cable, centrally shared modems, one common data base and a single management discipline. (See Figure 15-1)

Expectations of the intelligent premise controller go beyond the initial integration of the three systems just mentioned. To meet varied needs of individual tenant users, building developers will seek a system that will support circuit switched asynchronous or synchronous data up to 56 kbps; have a universal, two-pair wiring scheme to facilitate mobility; and provide adequate horsepower to support the three applications in one system and to add future applications.

To support three networks in one system, redundancy is a sought-after feature, with two processors, two disk drives and duplicated memory capabilities. Most important, as a safeguard for future growth within tenant organizations, non-blocking is a fundamental asset. Non-blocking capabilities assure that if everyone in the tenant environment wanted to go off-hook on the telephone or on-line with the data terminal during the same time period, a connection would be available for each. By combining the outstanding features of the PBX and the local area network all into one system, total communication reports are provided -- given that the call detail recording information is present for both voice and data switching through the system.

From the standpoint of an STS provider, a migration path of upward mobility is necessary for planning to accommodate tenant expansion in the future. Since in the shared-resources environment of STS (Shared Tenant Services), a single building can house hundreds or thousands of users, the integrated premise controller can grow in size, supporting digital as well as analog instruments. These expectations, or considerations, should be present in order to integrate a local area network into a premise controller. In the tenant environment, voice connectivity is the baseline. Value-added services including data switching and local area networking differentiate the facility's offerings, enhance the building, satisfy the tenant population, and provide profit and long-term viability for the STS provider.

Foundation for Expectation

Just as a building must have a proper foundation to support its construction materials, a foundation of state-of-the-art technology is necessary to uphold the expectations of desired applications in a communications systems. Those applications, in this case, are voice/data switching and local area network switching. To accommodate the growth requirements, factors such as non-blocking and large memory capacity should be included with large-sized 32-bit processors. They should be capable of supporting up to 10 megabit packet switched data speeds, and applications such as Ethernet or IBM coax replacement.

With the integrated premise controller in place, all STS users' applications can be supported - voice, data and local area networking - all across two twisted pair wiring and all transparent to the user. So when a new tenant arrives and purchases office space, not only are his telephone and data requirements met, he also has access to a multi-vendor world supported by a single, non-denominational system with one common data base and one cabling scheme.

The intelligent premise controller allows data access to the tenant whether he subscribes to it or not. If migration towards data is chosen, the user can simply plug into the system, rather than necessitating the developer or STS provider to return, tear out the walls to install another system and disrupt the business environment. Over time, cost savings can be substantial both to the STS provider and to the tenant.

The question may arise as to whether or not a great deal of horsepower is necessary in the early stages of a building's communications life. Anticipating future requirements, one must remember that it usually is possible to save half the costs by installing a powerful system initially - even if data access isn't immediately desired or the building is not fully occupied - rather than to return in a few years and string another wiring plan to add data and additional desired features.

As another example of cost, some industry figures indicate that the 3278 IBM terminal is moved five times in its lifetime. Depending upon the geographical location and the particular maintenance facilities a tenant has available, the cost ranges between $500 and $1,200 to move a coaxial cable for a terminal. These figures vary for each individual company, both in the number of times moved and in the cost for the move.

A company installing an integrated premise controller generally places two twisted-pair wiring everywhere the possibility of having a telephone or terminal may exist because of the wiring's low installation costs. Therefore, when a tenant user needs to move his terminal, no new coax cable has to be installed. He simply picks up his terminal and his telephone, carries them to the new location, and plugs them into the wall. After one data base change into the system, he is ready to operate again. Thus, the time of the moving user, soft dollar productivity costs is saved as well.

Common Ground

As demand dictates, tenants accumulate a number of terminals and computers for varied applications. Oftentimes, these devices are dissimilar in that they speak different protocols, or languages. Later, with different needs, tenants determine the proposed value of individual and incompatible systems being able to communicate with each other.

For example, the engineering department has a system. Marketing and accounting each have their own systems as well. Eventually, engineering may need to share data information with marketing; accounting wants to discuss materials purchased; and the search begins for a method to link these dissimilar devices.

To facilitate communication among the equipment, format or protocol conversion, devices are integrated through the intelligent premise controller to translate the data languages and to access diverse external networks transparent to the tenant user. With an intelligent premise controller in the building, established with the proper "info-structure" or growth and networking capacity, multi-vendor access is granted. The attachment can be made for touch tone telephones, integrated electronic telephones, ASCII terminals, office environmental apparatus, sophisticated workstations, IBM 3270-type devices, Ethernet equipment and future capabilities which have not yet been announced.

As an illustration, by utilizing protocol converters, tenants' asynchronous data devices can access X.25 packet and international data networks.

The integrated premise controller can be viewed, in a sense, as having communications tentacles extending to all locations of a tenant environment which require voice and data capabilities. By providing information channels to each user, the integrated premise controller becomes the hub of the automated office. Therefore, it creates the most logical means to interface with machines tenants employ to implement these desired tasks and to provide access to multiple databases. Through this integration, connectivity to 3270 networks becomes available for compatible communications among multi-vendor terminals, printers and computers from sources like DEC, VAX, IBM, Wang, Hewlett-Packard and others.

PBX-BASED LAN VERSUS STAND-ALONE LAN

The coexistence of a PBX and a stand-alone local area network is possible; but in such a circumstance, limitations are placed on the local area network. For example, with a stand-alone local area network, terminals A, B and C are connected within the network by coaxial cable. Each has a specific line allocated to it at the host computer so that it can access the network at any time. In order to add another device to the network or to move ones already established, it is necessary to install additional coax. The new coax for the move must be placed either in allocated space under the floor or through the ceiling. A database change would be a requirement, as would wiring adjustments when attempting to access varied host systems. When using the LAN from remote locations, a modem would be necessary.

With an integrated premise controller-based LAN, coax elimination is achieved through the use of twisted-pair wiring which results in cost savings when devices are installed or moved.

In this example, the LAN is accessed through the tenant's telephone, dialing to a group of lines on the integrated premise controller. Although space is always available for the user, he contends for any one of the group of lines with other users, rather than having one specific, assigned line as with a stand-alone LAN.

The integrated premise controller-based LAN gives the appearance of channel attachment which allows more flexibility for the system's configuration within a specified geographic location. When moving devices, installing new wiring is not required. By using twisted pair, the conduits are already in place from the initial installation of telephones. Similarly, access to varied host computers within a system is achieved without the cost or the labor of a wiring change.

Certainly, the integrated premise controller of the eighties and beyond is to be much more than a traditional PBX. It will localize the LAN's much more than the purveyors of local area networks might imagine, into a truly departmentalized network. By doing so, integrated premise controllers will become the stitching that brings together the fabric of the small tenant's LAN with that of the more pervasive communications link that is building-wide. It appears that the integrated premise controller will make the LAN more localized and more specialized in very high bandwidth situations. The intelligent premise controller enables a LAN to become the interface for a mainframe, thereby providing service to the large tenant with mainframe needs while also serving several small Ethernet local area networks through the same wiring system. Partitioning capabilities of the integrated premise controller simultaneously ensure privacy and yet access to shared resources.

A misconception about an LAN is that once installed, it is the panacea to which everything will attach. That's not exactly the case. For a particular device to be attached to a local area network, an interface must exist. This takes many forms to propagate the desired information from its point of origin out to another type of interface device on a local area network to the end terminal. Basically, an interface has the connector that a terminal needs, along with the protocols and link-level protocols that a particular terminal normally speaks. Therefore, the system is actually identical to the other networks in the fact that once the end interface is provided, the information can be disseminated through the system.

ENTERPRISE AUTOMATION

The thrust in modern business technology has been towards office automation, and rightfully so to this point. However, the concept of office automation is somewhat limited in scope. To truly automate an enterprise, which goes beyond the boundaries of departments or offices, one must examine information and its flow paths: in factories, in departments and outside the company. For example, when creating a new product, information may travel among product management, engineering, marketing, sales, customers, the media and more. Information most often travels outside the boundaries of an individual office.

The idea of enterprise automation established a foundation for accessing and communicating information regardless of where it may be or in what form. With the integrated premise controller as the backbone of the principle, geographic and enterprise boundaries are lifted.

To illustrate: a company's engineering department may utilize CAD/CAM systems; finance may have PC's linked to an LAN, and the factory may have robotics connected through a token ring network -- all tailored to meet the specific needs of the individual departments. The integrated premise controller serves as the communications gateway by facilitating the communication between these work groups through format and protocol conversion. Similarly, in enterprise automation the integrated premise controller provides more pervasive services which address applications that are common to all departments as well, such as voice and text messaging in addition to voice and data communications.

Information knows no boundaries; enterprise automation is the key towards erasing boundaries, thereby maximizing work group interaction and processing.

WHAT LIES AHEAD

The future retains an element of uncertainty in every field of business. Telecommunications is no different. The most practical tactic for strategizing is the study of the integrated premise controller's "info-structure." When calculating expectations, questions to ask include: is the percentage of bandwidth allocated for expansion adequate? is the integrated premise controller capable of supporting industry standards currently under examination for acceptance in the years ahead, for example ISDN and IEEE?

Since such standards have not yet completely evolved, an integrated premise controller cannot actually be compatible with them; however, it can be capable of adapting. To illustrate: ISDN is expected to be 2B + D signaling at the desktop, which is two 64 kbps timeslots, or 128 kbps, plus an additional 16 kbps timeslot for signalling, for a total of 144 kbps. Therefore, under the premise that ISDN is to evolve, if a building's communications controller has at least 144 kbps in place, then that existing bandwidth may be tailored towards an ISDN standard.

SUMMARY

The concept of data connectivity via integrated premise controller and local area network integration is becoming a reality. Providing universal connectivity, a must for future minded tenants, the integrated premise controller enables STS providers to network their communications operations with those of other terminals and computers. The cost - and the labor - of layering a building with individual systems for each feature can be avoided by the STS provider who recognizes expansion probabilities of tenants and the importance of readily available new communications capabilities.

Chapter 16

Using Digital Telephone Techniques For Building Automation Communications

Carl Pederson and Dennis Miller
Johnson Controls, Inc.

The use of telephone wires as a highway for communication between modes of computerized building management systems has long been recognized as a potential means of reducing installation costs. Integrated Services Digital Networks (ISDN) will enhance the telephone media by augmenting data transmission rates and multiplexing techniques while providing an expanding service to the voice telephone user.

Additionally, the advent of Local Area Networks (LAN) with their capability to interconnect many pieces of dissimilar digitally oriented equipment via a compatible communications interface, has given rise to a totally new concept in building services. This concept is referred to as the Smart or Intelligent Building.

Services referred to in the Intelligent Building include such things as electronic mail, word processing, data base storage, telephone service and billing, security service (badge access), fire detection and alarm, and many others. An additional capability will be the ability of the ISDN data highway to furnish the communication needs for building management services. These include comfort management, energy management, fire and security management, and miscellaneous services associated with building management.

This chapter will discuss the communication needs of systems which provide building management services with the

Fig. 16-1 Digital PABX

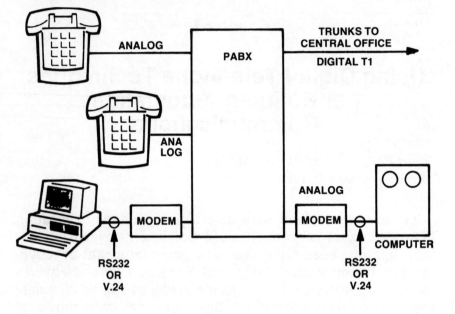

capabilities of an ISDN implemented telephone system. It will then conclude with several scenarios which illustrate the use of ISDN in conjunction with building management systems.[1]

Integrated Service Digital Network

Integrated Service Digital Network is a specialized form of Local Area Network. When used in conjunction with a digital switching unit, is capable of providing telephone service simultaneously with high speed digital data transmission including discrete addressing. The switch can be a local telephone company offering or a resident PBX which is specifically configured to support ISDN.

Digital transmission of voice was introduced in the early 1960's. This development permitted the simultaneous transmission of up to 24 voice messages one way over a single twisted pair copper line. With regeneration, the voice data can be retransmitted without loss of fidelity indefinitely thus permitting a virtually unlimited transmission range. This transmission system is known as T1 Carrier.

The ability to digitize the audio telephone information and the advent of micro processors provided the incentive for the development of the digital telephone switch. All of the modern PBX units rely on digital switching even though the lines to telephone subscribers are still analog. Figure 16-1 shows the basic configuration of the digital switch PBX.

An important element of a digital PBX is the Line Card. This card is capable of terminating multiple analog subscriber lines and converting the analog information into its T1 compatible digital equivalent. The culmination of T1 Carrier and digital telephone switch has led to the concept of an Integrated Services Digital Network or ISDN. Figure 16-2 depicts an ISDN - PBX.

---

[1]Building management systems are also termed energy management systems (EMS) and building automation systems (BAS). In the discussion which follows the latter is used.

Fig. 16-2 ISDN PABX

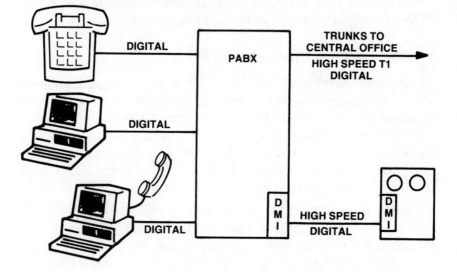

This concept is highly flexible as to destination switching and circuit configuration, and provides end-to-end digital connectivity. Three separate digital channels are available to the user. Each one can be individually addressed and configured. Two of these channels operate at a 64 kilobits per second rate and are referred to as "B" channels. The third channel is the "D" channel and its data rate is 16 kilobits per second. Standards are being decided upon for the control and access of these channels.

The standards by which an ISDN will operate are global in scope and are set by the Consultative Committee International Telegraph and Telephone (CCITT). Basically the CCITT defines the interface configurations and the protocol to be adhered to for various segments of an ISDN system.

Communication In A Building Automation System

To see how ISDN may be utilized in Building Automation Systems the BAS is renewed with emphasis placed on protocol, data rate, distance, media, message content, and supervision. This is applicable at both the headend computer to field processing unit (FPU) link and the FPU to sensor/actuator link.

The media most used throughout the BAS is 18ga shielded twisted pair. These lines have a 1 db loss characteristic for each 1000 ft. of length. In a particular application, the transmitter/ receiver cards are capable of operating with a 30 db attenuation. Since the end of line FPU must have a terminating resistor, a three db loss element, the maximum line distance figures out to be (30-3)x1000 or 27000 ft. for a point to point circuit. A multidrop circuit configuration is standard for many BAS installations. Many combinations of branch and string elements are acceptable so long as the db budget is not violated. Figure 16-3 depicts a representative installation.

The various items of BAS equipment that can be interconnected on a BAS trunk and will work with the headend equipment are:

Fig. 16-3 Generic BAS System

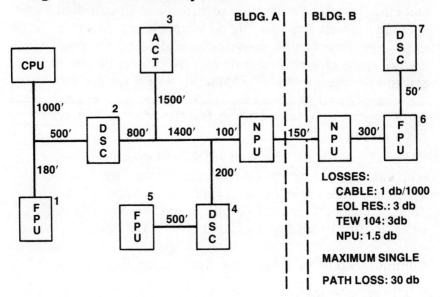

1. FPU - Field Processing Unit (-3db)

2. DSC - Digital System Controller (-3db)

3. ACT - Access Control Terminal (-3db)

Three other devices should be considered in a BAS communications line.

1. TEM-104 - Line Splitter (-3db)

2. NPU - Network Protection Unit (-1.5db)

3. EOL - End Of Line Resistor (-3db)

The line splitter is required on systems which require communication line supervision. This isolates the various bridged on circuits from the main line permitting complete supervision of those portions of the circuit which carry sensitive signals. The line splitter contributes 3db loss applied to the splitter leg only.

The NPU protects the line and its connected equipment from lightning and in some cases man made high voltage contamination.

The location and number of NPUs depends on the configuration of the line and its vulnurability. A 1.5 db loss is contributed by each protector.

The EOL - End Of Line resistor -- is required on a branch circuit in excess of 200 feet or at the end of the branch circuit farthest from the headend.

The supervision issue deserves close scrutiny because ISDN fails to provide this capability. Supervision is generally required when fire and security are incorporated in a system. This requirement is fulfilled by virtue of each of the end units so implemented to sense an open in the twisted pair circuit supplying that unit. A DC voltage is impressed on the pair at the terminal unit and a resistor is bridged across the far end of the circuit, thus permitting a current to flow.

The terminal senses this current and even though there may be a break in the circuit which will permit the BAS normal signals to still get thru via capacitance coupling, the absense of DC line current triggers a binary card in the terminal to send an alarm message to the headend thus signifying a line fault. Each branch of the circuit must be implemented separately so that a fault in one branch is not confused with one from another one, or is masked by other DC currents.

The separation is achieved by the use of isolating line transformers. Those branches which do not contain sensitive points such as fire or security sensors do not need the isolation devices.

The signalling method employed over the twisted pair information trunks of the example system is frequency shift keying with the outbound signals (headend to point modules), 45 kHz, a logical 1, and 39 kHz, a logical 0. The inbound signals from the field gear to the headend are coded with 18 kHz and 12 kHz frequency shifted to represent logical 1's and 0's respectively. The data rate in both directions is 9.6 Kbits/s.

Message length ranges from one, eight bit byte to 128 eight bit bytes, depending on the function. A dead time of approximately 20 ms. is inserted between each poll at the headend to provide a window for field replies.

For purposes of this discussion, it is not necessary to delve into the fine points of the message format or makeup. It is; however, important to understand is the sequence of conversation between the headend of the system and the various field entities.

The headend is the only originator of messages. All slave units connected to the system only read the headend's transmission, while in turn the headend reads all messages put out by the slaves. If a slave location fails to answer a pole, the master attempts to repole the slave three times. If the master is still unable to raise the slave, it will then record that slave as being offline and take appropriate action. This protocol provides for an orderly exhchange of information with the slave. Replying messages are interleaved between the masters actions. So long as the system is operating on a proprietary communications system, the large amount of two way traffic is effectively handled by virtue of the design.

BAS Involvement With PBX Related Systems

At present, there are already Building Automation System (BAS) products which use spare line pairs frequently found in PBX telephone subscriber line wiring cables. Those PBX/BAS Links are restricted to this source of communication path, but can be adapted to most available pairs so long as they do not exceed the maximum length or are unduly noisy.

Communication over a PBX/Link is accomplished by a voltage level scheme and may be implemented in either of two configurations; a two wire half duplex or a four wire full duplex.

The data transferred over a PBX/BAS Link interconnect has a density of 9.6 kilobits per second which applies to both the full as well as the half duplex.

Transmission distances possible over the PBX/BAS Link is dependent on the mode of operation, half of full duplex, and the number of drops on a particular line. A full duplex point to point connection can be extended to 7000 feet. The half duplex arrangement is limited to 4000 feet maximum. The allowable transmission distance on a four wire circuit for a maximum of 5 devices decreases to 3000 feet. A 500 foot cable length is allowed between drops which are daisy chained.

The data placed on the twisted pair medium used in the PBX/BAS Link implementation is base band in nature. An 8 volt positive potential is impressed on the two wire system by the headend terminal. This voltage signifies to the drops that the system is active. For transmission from the headend to the drops, the intelligence is transmitted via a positive 2 volt logical 1 and a negative 5.5 volt logical 0. The return signals from the drop are generated by the drops line interface causing a variable loading of the impressed line voltage, 7 to 5.7 volts for a logical 0 and 1 to 0 volts to represent a logical 1.

Fig. 16-4 ISDN Channel Division

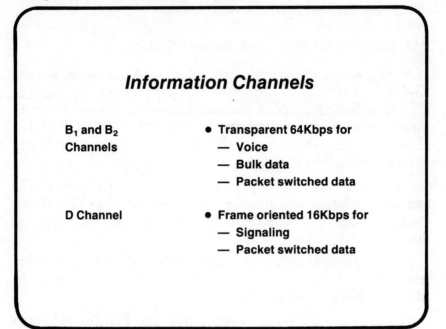

The full duplex case, since the lines are powered at the sending end for both the master station as well as the drop, uses a more conventional keying method. For both directions in order to send a logical 1, a negative 1.6 volts is impressed on the line. Transmission of a logical 0 requires that a positive 1.6 volts be sent.

The circuit interconnection at the headend is provided by a trunk termination unit. This unit interfaces directly with a BAS headend. For other than a point to point system, a networking box is employed. This box will permit up to 16 lines to be serviced from one headend. The lines radiating out of the networking box, in the case of the half duplex configuration, can support one FPU or DSC terminal. The full duplex situation permits up to five devices to be connected in a daisy chain fashion to each line.

The PBX/BAS Link uses existing building wiring for the most part to convey its intelligence. It's intent is to use spare telephone lines which exist on most buildings.

ISDN Characteristics

ISDN standards are international in scope and are set by the International Telegraph and Telephone Consultative Committee (CCITT). The CCITT defines interface configurations and the protocol for various segments of an ISDN system. ISDN provides for digital transport access at the telephone subscriber location. For the most part, this includes all areas of new and existing buildings. This service is to be completely flexible as to destination switching and circuit configuration and provides end to end connectivity. Three separate digital channels are available to the user. Each one can be separately addressed and configured. Two of these channels operate at a 64 kilobit per second rate and are referred to as "B" channels. The third channel is the "D" channel and its data rate is 16 kilobits per second. (Figure 16-4)

There are four basic interfaces involved in ISDN. They are the S,R,T, and U interfaces. At this writing the S,R, and T specifications are substantially agreed upon. The U interface is still in committee. The four interfaces are catagorized as follows:

S - ISDN end terminal interface.

R - Non ISDN end terminal interface to an S interface.

T - Terminal interface, similar in character to the S interface, but used between the switch or concentrator and the U interface.

U - Local distribution to central office interface.

The S interface is characterized by the following features:

1. Capable of 2B + 1D Data Rates

2. 192 Kbps Full Duplex

3. Optional Remote Power Feed

4. Pseudo - Ternary line coding

5. Up to Eight End Terminals Per Circuit

6. 1000 Meters Max. Line Length

7. Four Wire (2 pairs)

8. LAP D Protocol on D Channel

9. X.25 Protocol on One B Channel

10. Collision Protection Scheme

The R interface is characterized by the following features:

1. Will accept RS232C, RS422, RS423, V.24, and V.25 etc. conventions

2. Up to 19.2 kilobit per second data rate

3. Asynchronous to synchronous conversion

4. S interface compatibility with bit stuffing to round out frames to 64Kbts

The T interface is characterized by the following features:

1. Same operating parameters as the S interface

2. Can't be used as an orginator interface

3. Used to connect a transmission line (NT1) to a multiplexer or PBX (NT2)

The U interface is characterized by the following features:

1. May be two or four wire full duplex

2. Data channels min. 2B + D to max. 31B + D

3. Transmission distance not specified

4. Signals capable of regeneration

5. Several transmission schemes have been proposed, echo cancelling, T1, and ping pong

Figure 16-5 depicts the relationship of the four interfaces and Figure 16-6 expands on the channels and addressing scheme.

Present BAS Using ISDN

A hypothetical BAS control and sensing layout depicted in Figure 16-3 is typical in its configuration and equipment make up. In this circuit, the CPU or headend does all of the interrogationof the field equipment. The outlying units only reply to interrogations and never originate a transmission. Intelligence is transported over a shielded twisted pair of wires which has a typical lines loss of 1 db. per 1000 feet of length. Certain rules apply concerning line losses and when all is said and done, the total loss for a circuit cannot exceed 30 db. As noted in this example, all of the field units are strung out in a single circuit multidrop configuration.

Data rates applicable to this BAS generally do not exceed 9.6 kbit/s and each interrogation is accomplished in about one millisecond.

The master or headend then waits for 20 milliseconds for a reply from the interrogated field unit. This is also a one millisecond burst. If no traffic is forthcoming either to or from the field unit, the headend then interrogates the next field unit in line, and so on. If there is traffic, there is an exchange between the selected field unit and the headend. This protocol is

Fig. 16-5 ISDN Extended Architecture

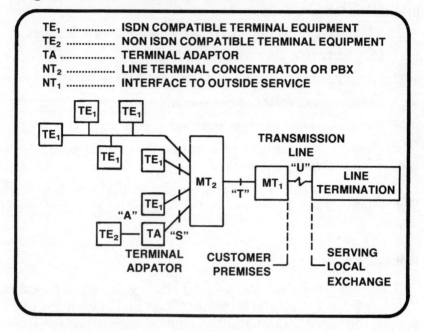

Fig. 16-6 ISDN Channels And Addressing

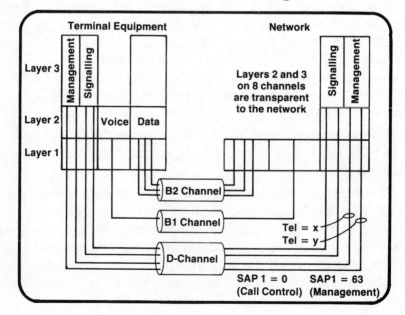

continuous and in the event of a no answer from a particular field unit, after three tries to establish communications, that unit is bypassed and an out of service message is generated.

In order to evaluate the potential use of ISDN as the BAS communication media, a number of characteristics of both systems must be considered. First, it is necessary to replace the multidrop configuration with a home run set up. This is because the system is also interested in the use of the lines for telephone service.

An important consideration are the distance restrictions placed on an ISDN multidrop configuration both in the absolute distance of the run, and in the configuration of the circuit. A provision in the ISDN configuration does not permit intercommunication between terminals which are connected to the same S interface circuit. Therefore, the telephones on these drops are prevented from communicating with each other. Both the distance and drop characteristics of ISDN are restrictive to BAS application.

It is possible to circumvent the limitations of the S interface. The use of the U interface back to back with S overruns this distance limitation. Another answer to this problem would make use of the LX (Local Exchange) at a strategic location, thus reducing the S interface lines to an allowable length.

Operationally, the ISDN interconnected system could be made to function in a satisfactory manner. The headend unit would be required to generate suitable addresses so that the PBX could correctly route each call to the proper field termination. The call rate to be comparable to the example system would be 50 interrogations per second. The PBX has the capability of routing a call through the switch in 20 to 50 microseconds, therefore, plenty of time to process the controller call rate.

Savings might be realized in the wiring and installation costs when substituting ISDN for a proprietary shielded twisted pair service. To offset this, the terminal requirement, a Terminal Adapter (TA) for each field unit would detract significantly from the wiring savings. One positive characteristic is the flexibility afforded by the ISDN approach. There are few restrictions on field gear location within the prescribed distances, so long as there is a phone location close by.

This scenario which couples the ISDN to a conventional BAS system illustrates the following:

1. Potential for reduced wiring cost

2. Integration of BAS functions with a voice system

3. Potential simplification of field buriel location

4. Potential for greater ease of access to remotefacilities

5. Potential problems in the use of polling protocols

6. Wiring cost savings could be offset by added terminal costs

7. ISDN line runs would be shorter when compared to current BAS maximum distance allowance. The above summary indicates that ISDN and BAS interfacing can be accomplished, however, there are potential problems with such an interface.

ISDN And BAS/HVAC

An alternate way to integrate ISDN and BAS functions is to use the ISDN's "S" interface. Figure 16-7 is a conceptual diagram of an ISDN/BAS. The 2B and 1D channel availability will permit the flow of necessary control information between the various BAS modes (part time D channel usage), medium high speed data communications on one B channel and regular phone service on the other B channel. The D channel will also be used to instruct the PBX as to the desired address through the switch for each message.

The modern telephone service via a PBX in a commercial, educational, or industrial building environment can provide telephone service via twisted pair circuits to every area of a building complex. All of these pairs are direct lines from the switch to the subscriber's telephone set. In most cases, the wire distance of each run is less than one thousand meters.

ISDN specifications limit the maximum length of a single S interface circuit to 1,000 meters. Although it is possible to multidrop TE1 ISDN terminals on a single S interface, there can be no communications between the terminals due to operational requirements. In addition, the allowable line length decreases rapidly with multiple TE1's on a S interface. (See Figure 16-8)

In BAS communications there are both packet type messages and virtual circuit requirements. ISDN will handle each kind with equal ease. The benefit of this technology lies in its flexibility. A particular path through the switch need exist only so long as the end to end connectivity is required. Once the job is finished, the circuit is broken and the various component parts are available for other services.

In order to realize the full potential of ISDN as the data highway for a Building Automation System, and still permit the PBX to furnish all of the services the modern communication user has become accustomed, the BAS master slave mode of operation must be replaced with an all master system. The concept of polling would overload the PBX when its demands are added to the normal telephone activity. The most effective system would communicate only when a particular BAS node has something to say.

Circuit assurance would be maintained via a routing service provided by the PBX. Some reasonable reporting frequency from each node would ensure that the nodes were indeed active.

An example exchange with the ISDN "S" interface for the hypothetical BAS system, might be as follows:

* D -- Packet to PBX - Send Dial Tone to B1 Channel
* B1 - Dial tone to TE1
* D -- Packet to PBX - Connect B2 to File Server
* B1 - Dialing Signals to PBX
* D -- Packet to TE1 - B2 Connected to File Server
* B2 - File Data Signals Online

 * D -- Packet to Fire Panel and Outside World - File in
    Area 21
 * D -- Packet to Fire Engine - HVAC Damper Closed
 * B1 - Telephone Connection Active
 * D -- Packet to All Parties - Emergency Over

In the meantime, the data being transferred over the B2 channel continues without interruption until the job is completed. As with the PBX systems in general use today, what is happening on one connection through the switch does not influence operation of the rest of the switch.

This ISDN/BAS marriage has more benefits than the obvious ones so far alluded to. First of all, due to the extensive installed plant of telephone wiring within a building, the inclusion of the BAS onto the phone system permits a high degree of flexibility in placement of the field units.

In this scenario, the sensors and actuators are not ISDN implemented. Most of the signalling between the sensors and actuators is DC., or very low frequency analog AC., which would not be economical for conversion into the ISDN network. (See Figure 16-8) These runs are short, typically under 50 feet, and are better suited to other forms of communication.

Through the utilization of the B channel of the ISDN, high speed information transfer between elements of the BAS is easily accomplished. This information channel can be set up either as a virtual circuit in which batch transfer of data could take place, or in packet switch mode for the handling of short, to the point messages. This service would operate at a data rate of 64 kilobits per second. Just like the D channel operation, the B channel is equally flexible.

The BAS with ISDN/PBX communication is able to communicate with the outside world because of the U interface standards generated by CCITT. This feature will enable one BAS to be linked to other BAS sites over the switched telephone network. It will be just as easy for an operations manager in Milwaukee, for example, to access a data log from the BAS controlled building location in San Francisco as it would to get local data.

The PBX capable of supporting an ISDN is in itself computer based, and as such, capable of functions in addition to operational chores of running the switch. This processing power along with the associated memory could be utilized in performing BAS tasks such as energy management, coordinated comfort control, fire and security inter-relationships, and facilities management. The ISDN would act as the interface which would provide the necessary communications between the switch based processor and the outlying field gear.

For BAS equipment that is not ISDN compatible, the CCITT R interface will permit these non ISDN data formats and rates to be accepted into the ISDN environment. Also, for distances which exceed the 1000 meter limitation of the S interface, the U interface can be employed as a line extender.

Conclusion

When will this metamorphisis take place? The technology necessary to make a reality of ISDN exists at this time. The two stumbling blocks that must be resolved before ISDN will be accepted are; finalization and acceptance of the standards developed by the CCITT, and the availability of low cost VLSI parts which will accommodate the ISDN interfaces. There are three general groups of estimates as to when ISDN will be as commercial reality. Some say that there will be offerings by the late 1980's. The majority opinion is that there will be significant ISDN product in the marketplace in the early 1990's. The most pessimistic pundants forecast a 1995 timing.

The most important fact is that within the next 10 years ISDN will be a realistic fact of life and that the capabilities and advantages of ISDN will provide increased functionality along with the potential for reduced installed cost for services such as BAS.

Modern technology makes ISDN possible, as well as financially practical.

The ISDN evolution will change the relationship between the computer and communications industries. Communications is no longer possible without computers; computers finding their real value when they can communicate. Over the next decade, the evolution of ISDNs in the United States is expected to unfold during three consecutive phases: As seen in Table 1, these phases are transition, first generation, and second generation.

The transition phase, in which the industry currently exists, is distinguished by common channel signalling, as well as digital capability in local loops and switching exchanges.

Moreover, increased use of pre-ISDN services, such as AT&T's Local Data Transport and Circuit Switched Digital Capability, will be seen.

The CCITT Plenary Assembly was convened in November 1984 to produce the first ISDN standards. Completion of these standards will mark the climax of the transition phase and establish solid footing from which the first-generation can proceed.

The first-generation phase began in 1986 and will end in about 1990. Emerging in this period will be integrated access capability and CCITT-standard ISDN equipment and interfaces. Integrated access implies a single interface for a wide range of voice and data services, unlike the use of separate hookups required during the transition phase. Customers will have expanded control of the networks through D-channel signalling. For example, these customers will be afforded dynamic allocation of the bandwidths to which they subscribe (16, 80, 144 K-bit/s, and so on).

The second-generation phase is expected to start in 1990. Because the telecommunications industry is advancing so rapidly, it is difficult to predict precisely what this era will bring. But it is anticipated that a shift toward high-speed data and video capabilities will be seen, as well as the integration of circuit and packet switching into a single transport. Various new services are likely to arise as suppliers and subscribers alkike gain a better understanding of the capabilities and benefits of ISDNs.

TABLE 16-1 ISDN EVOLUTION

(Source Data Communications June 1984)

TRANSITION (1983-1986)

* PRE-ISDN SERVICES

* SEPARATE ACCESS FACILITIES - ALTERNATE
  VOICE AND DATA SWITCHING
  EXCHANGED

* INCREASED USE OF COMMON CHANNEL
  SIGNALLING

* 64-KBIT/S CLEAR CHANNEL TRANSMISSION

* CCITT ISDN STANDARDS

FIRST GENERATION (1986-1990)

* INTEGRATED ACCESS

* CCITT-STANDARD EQUIPMENT, INTERFACES

* SIMULTANEOUS VOICE AND DATA AT 64 KBIT/S

* EXPANDED CUSTOMER CONTROL - D CHANNEL
  SIGNALLING

SECOND GENERATION (1990 AND BEYOND)

* HIGH-SPEED DATA AND VIDEO CAPABILITY

* INTEGRATION OF CIRCUIT AND PACKET
  SWITCHING

* OTHER NEW SERVICES

Fig. 16-7 BAS/ISDN Conceptual Diagram

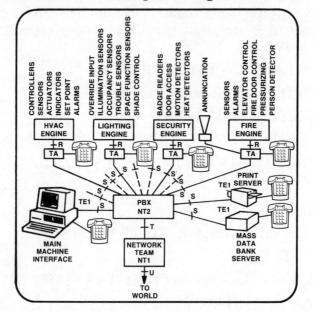

Fig. 16-8 BAS/HVAC Equipment Room

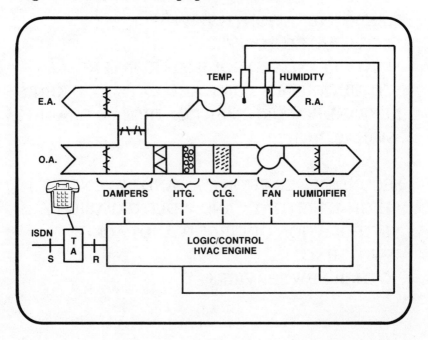

SECTION V

Shared Tenant Services

Chapter 17

A Tutorial Step-By-Step Implementation Of A Shared Telecommunications System

John Daly
Planning Research Corporation
Past President, Multi-Tenant Telecommunications Association

Much has been said and printed about Shared Telecommunications Systems (STS) and the Intelligent Building concept in the past two years. But very little effort has been made to concentrate the knowledge of the individuals who have actually been involved in the implementation of shared tenant services into one package. This tutorial is an attempt to begin that process but it is by no means complete or fully comprehensive. The information provided is based on a good measure of first hand experience gained as a result of having designed, implemented and managed the Planning Research Corp. (PRC) STS; and, because I have had the benefit of sharing experiences with my colleagues in the industry through Associations such as the MTTA (Multi-Tenant Telecommunications Association) and the IBI (Intelligent Building Institute). I recognize, however, that my perspective is limited. Like most sciences, the technology of Telecommunications and that of building automation and control systems as well as data processing and data management changes rapidly and a view expressed today may not be relative in a short period of time.

This tutorial is therefore addressed to the implementation of a shared telecommunications system, one that is typical of those that are in place and are functioning as saleable commodities. At this writing, approximately 250 have been installed and or in process of installation.

STS systems, access to word processing, data processing, data communications, reproduction equipment, Local Area Networks (LAN's) etc. are among the services that are most frequently mentioned in promotions as being essential to tenant needs. And I agree that these services are undoubtedly important to the conduct of any business. But, unfortunately, tenant demand for the features of this type of so called intelligent building is virtually non-existent at the present time. In PRC's experience, not one of the twenty tenants who now occupy space in our buildings came to us because they were aware of our capability to provide STS Service. The important factors were our location in McLean, Virginia, an area serving high tech business; our ideal proximity to the Federal Government marketplace, and our location equidistant to two airports. Jack Gavigan of Prudential Realty and a fellow Board member of Multi-Tenant Telecommunications Association (MTTA) put it rather succinctly in a recent letter addressed to me. He reported that in answer to the question from senior Prudential Realty management, "Does STS help to sell space?" he said "Not really, yet it will once tenants understand what STS is and start to demand it as a condition of their tenancy." The point that Jack made so eloquently in that letter is that real demand for STS and other Intelligent Building services can only be stimulated by educating prospective tenants through extensive promotions.

A percentage of commercial realtors and developers and some large corporations who employ professionals to find space are aware of STS and the Intelligent Building concept but they are moving slowly for a number of reasons. True familiarization with the operating efficiencies and a real appreciation of the economic benefits are basic stumbling blocks to their full support and participation. Additionally, these same professionals have difficulty in selling the concept to their management.

Hopefully, this tutorial will assist in the educational process by explaining in enough detail what steps must be taken in both the implementation and marketing of STS systems. I have organized it in the form of essential building blocks, each one a step to be taken as well as a requirement to be analyzed. Figure 17-1 offers a view of the elements and to appreciate each step, let's start with the Baseline, the Shared Telecommunications System.

Fig. 17-1 Shared Telecommunications System

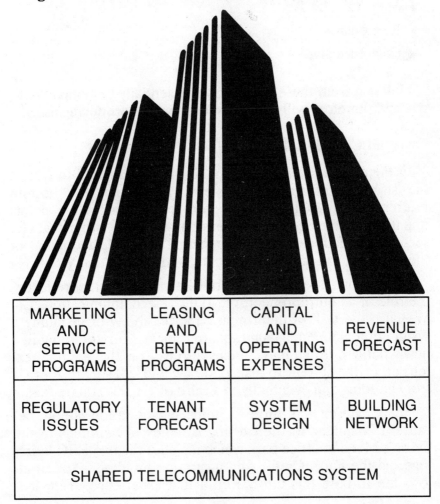

MARKETING AND SERVICE PROGRAMS	LEASING AND RENTAL PROGRAMS	CAPITAL AND OPERATING EXPENSES	REVENUE FORECAST
REGULATORY ISSUES	TENANT FORECAST	SYSTEM DESIGN	BUILDING NETWORK
SHARED TELECOMMUNICATIONS SYSTEM			

STS is not simply a PBX or a Centrex System. It is a <u>Service</u> accorded to tenant subscribers and it carries with it the basic requirement to answer a variety of business communications needs.

Above all, STS must be:

- Reliable
- Functional
- State Of The Art
- Responsive
- Cost Effective

Unless it is all these things it will essentially be noncompetitive and therefore will not become a revenue producing asset.

Essential Considerations

In Figure 17-1 we see the building blocks aligned to reflect the structure of the STS system. Those on the bottom reference the development of factors essential to the determination of the Economic Feasibility of the project. The second row delineates factors essential to the development of the business plan and the promotion and sales of the system.

For the purpose of this tutorial let's assume (theoretically) the role of an entrepreneur, the JDD Telecommunications Co. We have the task of determining the feasibility of installing a system in a planned 500,000 square foot, six story building in a newly defined office park. Plans also call for the eventual extension of the system to a second new 160,000 square foot six story building and eventually to a third of 130,000 square feet.

I have chosen this three building configuration simply because it is similar to the existing complex of buildings in the Tyson's McLean office park which includes the two PRC buildings and the Farm Credit Administration building. And, it allows me to base this tutorial on actual experience. But, it is not my plan to relate the Planning Research Corporation (PRC) case history. As the anchor tenant in its Headquarters building PRC established the first truly functional Shared Tenant System in 1981. Unlike many other STS systems, PRC service offerings included significantly more than POTS, (Plain Old Telephone Services.) But let's now continue to explore what must be considered to justify implementation of a conventional STS system in our hypothetical three building complex.

Regulatory And Legal Issues

Purpose: To understand the restraints imposed on STS revenue opportunities by the LEC tariffs and State regulations.

Based on my experience I can only advise that before any attempt is made to develop an STS system, a good attorney look into state and local regulations governing the STS service offerings. Prior to investing I would require written opinions on such questions as:

- Tariffs governing resale of local (intra-LATA) service
- Need to partition switches to prevent intercommunication between tenants and shared usage of local trunks, particularly those that are tariffed as Flat Rate
- Tariffs governing directory listings
- Termination of Local Exchange Company (LEC) cable facilities within the shared environment (Mean Point Of Presence-MPOP)
- Access by the LEC to serve tenants
- Rights of the LEC to prevent BY PASS
- Restriction as to scope of STS operation, i.e. to serve tenants on
    - Non-contiguous properties
    - Restrictions based on limitations to serve only tenants located in properties under common ownership or common management or both
    - Service warranties and guarantees to tenants

Fig. 17-2 Capital

- Construction costs

  - Switch room

  - Telephone equipment rooms

  - Environmental control equipment

  - Electrical service

  - Risers / conduit

- Equipment lease / purchase

  - Switch

  - Telephones / consoles

  - Power backup

  - Cable (building network)

  - Call detail recorder

  - Tenant software

  - Minicomputer (VOICE MAIL)

  - Miscellaneous

- Installation

  - Labor and material

Fig. 17-2 (continued) Operating Maintenance

- System (PBX) finance rates

  – Insurance and taxes

- Legal fees

  – Regulatory issues

  – Contracts and leases

- Consulting service

  – Economic feasibility

  – System design

- Marketing

  – Sales personnel

  – User training

  – Demo equipment

  – Brochures

Fig. 17-2 (continued) Operating And Maintenance

- Operating requirements

  - Space rental

  - Personnel

  - Office furniture / equipment

  - Telecom services

  - Electric power

  - Climate control

  - Supplies

- Maintenance

  - Personnel labor rates

  - Supplies / spare parts

- Administration

  - Billing

  - Receivables

  - Data base management

  - Bad debts

This last point is one that may not have been given the attention it deserves. PRC, for example, states in our tenant lease agreements that any and all warrantees provided by our vendors and the local telephone company are extended to the tenant subscriber. Nothing more is offered or implied.

After the legal and regulatory issues are explored and expensed, the next vital step is to assemble data essential to the determination of the economic feasibility of the project.

Economic Feasibility Study

Purpose: To determine the Revenue Potential of the STS business opportunity within a proposed environment.

The Economic Feasibility Study is second only in importance to the determination of the regulatory issues. It would make little sense to begin any investment process without first determining when a return on invested capital can be realized. STS systems are expensive and that expense starts immediately...right at the beginning of the planning stage. For example, in the course of resolving the regulatory and legal issues, an attorney's fee ranging upwards from $100.00 per hour is to be anticipated. And, of course, the time and material, travel expenses of those who are pursuing the STS business opportunity are also healthy start-up cost factors.

Basic Factors

Expense factors that are essential to the development of an Economic Feasibility Study are listed in Figure 17-2 and are not necessarily in the order in which they might occur. Legal, consulting and market development fees normally are the first to be incurred and invariably are costs chargeable to operating expense. They also represent expense items that must be taken into account each year in the life of the system. Allocation of funds for these activities as well as other operating expenses must be programmed over the life of the system and, particularly the period which the financed capital costs are amortized. Most systems today are depreciated over a period

Fig. 17-3 Tenant Participation Forecast

Building # 1 – 500,00 sq. ft.

Calendar year (end)	1987	1988	1989	1990	1991
Projected occupancy rate	35 %	60 %	80 %	85 %	90 %
Occupied footage (sq. ft.)	175,000	300,000	400,000	425,000	450,000
STS capture rate	20 %	40 %	60 %	80 %	80 %
Number of stations (@ 1 / 200 sq. ft.)	175	600	1200	1700	1800

of five, seven or ten years but, for the purpose of minimizing the extent of the data included in the Figures accompanying this tutorial, I have chosen five years as the initial life.

Five years is also an appropriate term for another very important forecast, Tenant Participation (or Capture Rate) Forecast. Figure 17-3 delineates a view of the number of station units that might be considered the CAPTURE RATE for an STS system serving a building of 500,000 square feet. But, the reader must take into consideration the fact that the Capture Rate percentage is purely hypothetical and does not apply to any known situation. What the exhibit matrix establishes, however, is the need to project a capture rate based on the expected occupancy rate during a specific time frame In Order To Size The System. Conversely, the determination of the initial system size and the incremental expansion over a period of years provides the information necessary to order and price the system. Figure 17-4 offers a forecast view of a PBX system configured on the basis of the capture rate delineated in Figure 17-3. This exercise also serves to establish criteria needed to develop the specification for the Building Network or as it is more often called the Premises Distribution System.

Premises Distribution System - Building Network

About the only service element installed initially which may meet the ultimate tenant need is the Riser Cable illustrated in Figure 17-5. In my view, this element of the Premises Distribution System (Building Network) should be designed to virtually meet double the requirement normally engineered for each station line, private (leased) line or special circuit.

For example, if the dimension of each floor of our hypothetical 500,000 square foot, six story building is approximately 83,0000 sq. ft., our need then is to provide cable pairs sufficient to serve a minimum of 417 station lines per floor if we use the conventional forecast of one phone/200 sq. ft. In this case we have 83,000 sq. ft. per floor divided by 200 and arrive at a conclusion of 417 voice terminals/floor. Considering that

Fig. 17-4 PBX System Size

Annual Requirements	1987	1988	1989	1990	1991
Ports	200	660	1305	1850	1955
Number of main stations	175	600	1200	1700	1800
Central office trunks	24	40	80	120	125
Long distance access lines	10	20	25	30	30
Standard 2500 type station	140	400	800	1200	1220
Electronic sets	35	200	400	500	575
Consoles	2	2	2	2	2
Voice mail / message center ports	–	–	–	–	–
Back-up power (AMP / HR)	–	–	–	–	–

Note: Data modules / ports not included in view

we will utilize 24 AWG, six pair station cable (an arbitrary decision for our hypothetical case) terminated in dual wall outlets or modular jacks in each work space or office location, as reflected in Figure 17-6, we should install at least 1200 pair of 24 AWG riser cable on each floor. And, as the probability exists that riser cable would be installed in two closets (one on each side of the building) a minimum of 600 pair should be cut down in each closet. Roughly, the installation of 1200 pair riser cable on each floor would insure the availability of two pair for each voice terminal and 400 pair for supplementary connections of data terminal using RS 232, 422 or 449 type interface arrangements.

As a supplement, or in some situations as an alternative, consideration should be given to the use of a fiber optic cabling system and/or multiplexed (e.g., T1) system where distance may impact signal transmission levels. In retrofit situations, because of limitations to riser channels, fibre or multiplex systems may be essential to overcoming space restrictions.

Station Cable

Our experience at PRC proves the viability of using 24 AWG wire for transmission of digital signals up to 4.27 MEGABIT/SEC. Currently we have connected literally dozens of 3270 IBM type and Wang VS terminals using Baluns (devices which provide matching impedance networks) thus eliminating the need for single coaxial or dual coaxial (WANG) cables (RG 59 or 62 AU). Care of course must be exercised to keep twisted pair station cable runs at least a minimum of 18" to 24" from other Electro-Magnetic Inductive (EMI) devices. Average distance for the connection of the Wang terminals to hosts using the 24 AWG twisted pair has been under the 500 feet specified by WANG. But, in contrast we have successfully used BALUNS at distances of 1600 feet to connect 3270 type terminals to their hosts. We have also connected 42 SYDIS (IVDT) terminals operating at 380 KB/s using two wire twisted 24 AWG at distances of 600 feet. Twisted pair, 24 AWG, will also serve ideally to carry the

Fig. 17-5 Cable Distribution Plan

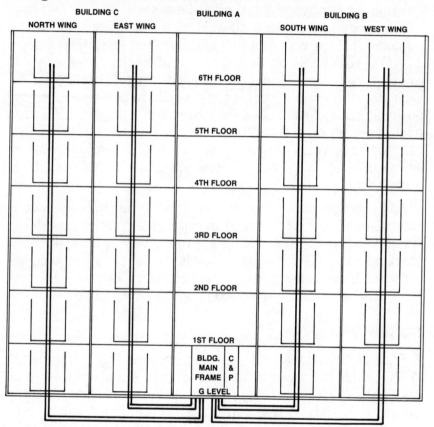

proposed Integrated Services Digital Network (ISDN) digital information bit steams of 64 KB/s & 1.5 MB/s.

For more information regarding the subject, I suggest reading Chapter 17, "Inside Wiring For The High Tech Building," which appears in High Tech Real Estate, a Dow Jones-Irwin publication edited by Alan Sugarman, Andrew Lipman and Robert F. Cushman. Chapter 9 of that book also covers the PRC-Case Study.

Wire Closets

As a result of constantly changing technology it is vital in the design of the Premises Distribution System to provide adequate space to house cable and wire terminals as well as interface equipment such as multiplexors, modems, circuit termination packages, etc. Backbone networks for Broadband and high speed Baseband (e.g., Ethernet) service also require adequate space Figure 17-7 presents a good schematic of a typical 10 MB/s Ethernet. Therefore, it is essential that architects and owner/operators provide spacious closets. In this hypothetical case situation I would recommend a minimum of 64 square feet per closet.

High Speed Networks

To provide an appreciation of a more sophisticated high speed Baseband network - a view of an Ethernet packet switched (CSMA/CD) 10 MB/s transmission facility is reflected in Figure 17-7. This network was installed in the PRC headquarters to provide a communications bus bar for the transfer of electronic mail between approximately 180 Personal Computers (PC's). Each of the PC's are hard wired to one of the eight ports available on Infotran multiplexors that are installed in the wire closets on each floor of the building. The muxes, in turn, are connected by transceivers to the 1/2" coaxial cable bus that is wired in a continuous loop between closets. The terminals, in addition to being able to address

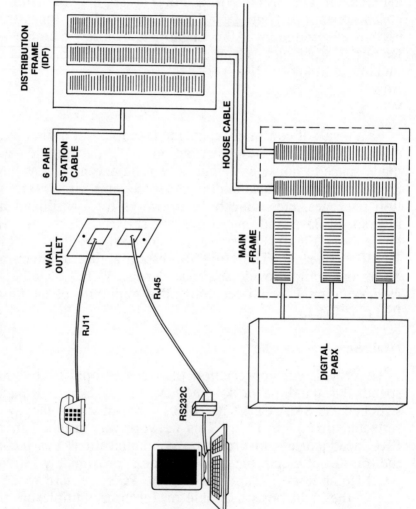

Fig. 17-6 Building Network

each other, have access to a DEC VAX 780, a Xerox Stars system and our main frame, a dual NAS 8000 system.

This type of Baseband facility or an equivalent Broadband using either coaxial cable or fiber-optic as the 'bus' may become a basic network in the intelligent building. For the near future, however, tenants having diverse business interests will be more likely to need access to the outside world. Shared modem pools accessible through the PBX should be considered, if the STS operator chooses to offer enhanced services.

Request For Proposal (RFP)

In determining the capital costs and the finance rates for the PBX system it is essential to develop an RFP that defines the system requirements and requests pricing from vendors for the initial first year's equipment requirement and the installation of the riser cabling system. Costs for incremental additions on an annual basis must also be requested along with labor, HVAC and power costs for each of the five years of the initial system life. Without question it is difficult to write an RFP to address a developing requirement; however, it can and should be done. In preparing the RFP particular attention must be given to basic STS system element costs such as:

- Redundancy
- Power Backup
- Tenant Software
- Call Detail Station Message Detail Recording (SMDR) Capability
- Traffic Handling Capacity
- Ease of Maintenance
- System Management

For more detailed direction as to writing RFP's I recommend that readers secure a copy of The BCR Manual Of PBX's written by Leo F. Goeller, Jr. and Jerry Goldstone.

Fig. 17-7 Information Resource Network
10 MBS Ethernet Coaxial Cable System

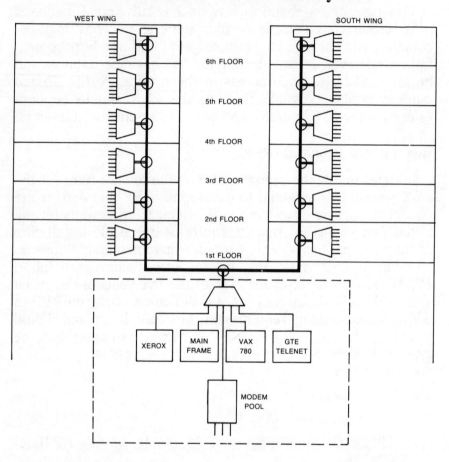

The book is a treasure house of PBX and Centrex data and contains 50 pages covering the subject of preparation of RFP's.

Selection Of The STS System

As there are a number of vendors in the marketplace today whose systems are specifically engineered to serve as STS vehicles, care must be given to select one that is modularly designed to accomodate inexpensive incremental additions over a period of the initial life (e.g., five years). Also, consideration should be given to the space the system is to occupy, power demand, and the environmental control expenses. And, as previously discussed, particular attention must be given to the selection of a system that minimizes the labor involved in the system administration tasks. As in any telecommunication system, the on-going task of managing moves, additions and rearrangements can be very expensive and time consuming if the Building Network is not well designed and/or the software provided in the switch proves difficult to work with.

Expense Projections

Assuming that careful evaluation of these criteria have resulted in the most appropriate STS selection, the next step is the determination of the construction and operating expenses. Figures 17-8 and 17-9 delineate the expense criteria to be projected over a five year initial life for the Capital and Operating system elements. But, it should be appreciated that the actual system life could be ten or more years. And, once the five year view is projected, it might be advisable to extend the economic feasibility exercise to ten or more years.

This cost evaluation process as well as the revenue projections can be readily accomplished on a PC or other microcomputer using a LOTUS, MULTI-PLAN or VISI-CALC spread sheet software program.

Fig. 17-8 Capital And Operating Expense Projection

	1987	1988	1989	1990	1991
Capital costs					
Switch					
Voice sets					
Data sets					
Message center / voice mail					
Riser cable install.					
Station wire install.					
Billing / SMDR EQP.					
Switch room / HVAC					
Total capital cost					
Cummulative capital					

Personnel Salaries And Benefit Expenses

Figure 17-10 provides a view of the factors to be considered as personnel expenses. For this hypotherical case I have chosen to delineate eight functions (staff positions) that would have to be filled to adequately serve a system of 2,500 lines or more. Expense factors for the marketing personnel (reflected in the exhibit) are addressed later in this tutorial.

Leasing and Rental Programs

As the basic revenue source is derived from the rental of station equipment and trunk or line ports it is essential to develop a program which provides the prospect with a contract for service which is competitive to that which may be offered by equipment vendors such as AT&T, unregulated Bell Operating Company units or other interconnect companies. Conventionally, STS operators offer fairly straight forward proposals to tenant prospects in order to avoid confusion and time consuming legal reviews. Figure 17-11 provides a delineation of the items most likely to be found in the Service and Equipment listing of a contract.

Another important contractural element to develop is the service term: From my view, rental of the service provided should be consistent with the term of the premises lease. If, for example, the tenant opts to sign a five year premises lease with the landlord, then the STS service should be offered on a concurrent basis. However, if the rental of the STS can be made more attractive, and/or competitive by extending the period for rental of the STS provided service, then it should be considered.

To protect from possible revenue loss should the tenant vacate earlier than the premises lease called for, it would be appropriate to include, in the contract, a penalty clause for early termination similar to that employed by other vendors or finance houses. Consideration might also be given to a buyout clause. Additionally, an alternative should be available, to

Fig. 17-8 (continued) Capital And Operating Expense Projection

	1987	1988	1989	1990	1991
Variable costs					
Consulting fees					
Legal fees					
L.D. trunk charges					
Local trunk charges					
Long distance cost					
Maintenance					
Bad debts					
Miscellanous					
Installation costs					
Subtotal					

Fig. 17-9 Operating Expenses

	1987	1988	1989	1990	1991
Space rental					
Office furniture / equipment					
Telephone and utilities					
Supplies					
Travel / entertainment					
Insurance / taxes					
Tenant billing expense					
Test equipment / tools					
Mail / messenger					
Training and education					
Computer / computer services					

Fig. 17-10 Personnel Salaries And Benefit Expenses

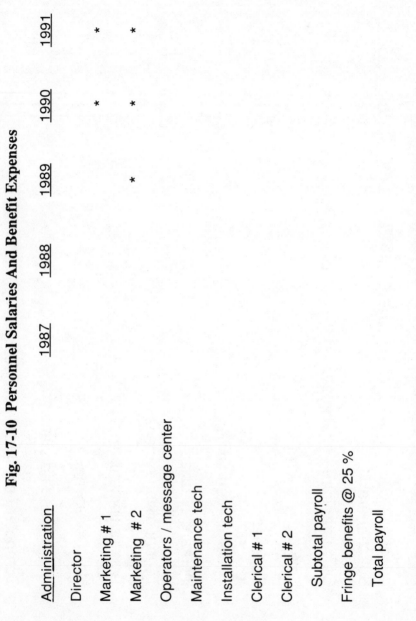

Administration	1987	1988	1989	1990	1991
Director					
Marketing # 1				*	*
Marketing # 2			*	*	*
Operators / message center					
Maintenance tech					
Installation tech					
Clerical # 1					
Clerical # 2					
Subtotal payroll					
Fringe benefits @ 25 %					
Total payroll					

* Indicates that the employment of personnel for this activity will be discontinued

allow tenants to purchase elements of the system rather than to rent. Indeed, although the basic PBX service (i.e., station and trunk/line ports) must inevitably be offered on a lease basis, the station equipment should be made available on a purchase basis. In either case, rental or purchase, equipment maintenance should be offered on a separately identifiable cost basis.

I have been asked on a number of occasions, if I would consider offering features (e.g., speed call, call forwarding, etc.) as a separately billable service. My answer has been no. As an administrator, I offer all the features available within the software of the switch and would not want to follow the unenviable posture of most BOC's in offering Centrex service on a piecemeal rate/feature basis.

Finally, I would make absolutely sure that all warranties are expressly identified and included in the contract. In essence, I would offer no liability for service interruption and would extend only those warranties offered by the LEC and the equipment manufacturers to the tenant.

Marketing And Service Programs

Assuming that we have cost data relative to capital investment, finance rates and operating expenses for all but the Marketing and Service functions we can think in terms of putting together a staff and setting up a marketing plan to educate and sell prospective tenants.

Figure 17-10 provides a view of the numbers of people that I feel are needed initially to accomplish the tasks involved in timely fashion. Note also, however, that I make an assumption that it will be unnecessary to carry two marketing personnel beyond the second year. And, that no marketing personnel need be carried on payroll beyond the third year. By that point in time I believe the capture rate should be nearly 80% of the occupancy rate (see Figure 17-3) and therefore ongoing marketing tasks can be assumed by the Project Director.

Fig. 17-11 Revenue Opportunities

Revenue sources	1987	1988	1989	1990	1991
Station and line connecting fees					
Station and line (port) rentals					
Station equipment rental:					
Single line (2500) instr.					
Electronic sets					
Data modules					
Data ports					
IVDT terminals					
Special assemblies					

	1987	1988	1989	1990	1991
Additional / moves / changes					
Direct dial numbering plan					
Pro-Rata share of trunk service					
Dedicated trunk / line port charges					
Long distance service charges					
Building / local area network access charges					
Message center services					

Recognizing that it is difficult to hire and develop good marketeers for a two year employment term it might make good sense to use a consulting organization to handle this aspect of the business. In addition to the development of a sales and marketing program, the consultant might also be charged with the task of estimating the occupancy and tenant capture rates.

As it is also essential to understand competitive offerings of the interconnects, the marketeer/consultant should be required to produce cost and benefit comparisons between the STS system operator's proposals and those that tenants are bound to receive from competitors.

In essence, a marketing and sales strategy has to be developed and put in place prior to start of leasing space. In fact, the availability of telephone service should be a feature included in the building leasing program. Realty agents handling the space rentals should be trained to at least introduce the concept and voice the benefits. The STS marketeer (or consultant) should then follow-up and provide a detailed proposal. An on-site demonstration facility should also be set up to allow the prospect to view the equipment. This on-site facility, in turn, can be used effectively to train tenants in the use of STS system features.

STS System-Revenue Opportunities

Depending on the regulations governing STS systems in each particular state in which the service is implemented, the STS administrator may offer either to provide a partitioned arrangement where the tenant agrees to pay for a certain number of trunks and extension lines. Or, as is the case in State if Virginia, the tenant can be offered shared Direct Inward Dial (DID) service as an alternative and charged a pro-rata share of the direct dial numbering plan and common trunk group expense. This expense would be in addition to the monthly rate for station lines and equipment installed for the tenant. Additionally, the tenant would be offered access to a Long Distance

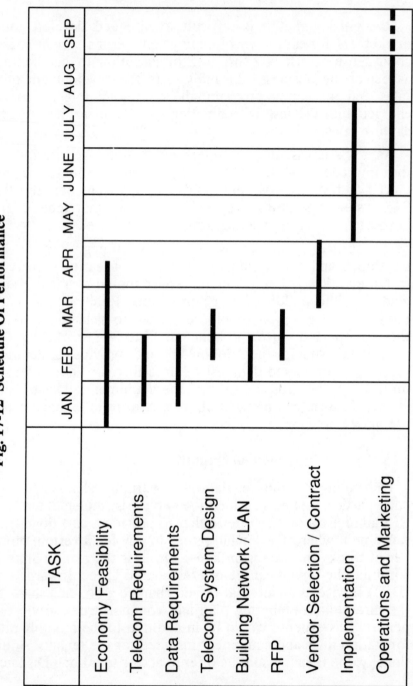

Fig. 17-12 Schedule Of Performance

(LD) network which may be directly provided and managed as an integral function of the STS system or, as an alternative, it might be provided by access line connection to one of the LD resellers switch nodes.

Local call charges (Message Units), because of State regulations, are normally a direct pass-through expense to the tenant. Until State regulatory bodies address the issue of reselling local service no surcharge should be applied to local message unit charges tariffed by the LEC (Local Exchange Company).

What is important to appreciate is that the economic feasibility of our hypothetical project is going to be determined on the basis of revenues derived from the rental of station and trunk ports and station equipment and from the resale of Long Distance.

To a lesser degree revenues are derived by charging for moving, changing and adding to established station and line configurations. Figure 17-11 lists these revenue sources, and as stated previously, the exhibit reflects the collection of these revenues over the five year initial term.

There are as yet no published statistics documenting the revenue potential of other services that may be provided such as access to word processing and/or other computer services. With these reservations in mind as to revenue potential of enhanced service lets move on to the next element.

Schedule of Performance

Figure 17-12 reflects a view of the Schedule of Performance which serves only as a reminder that it is essential to put time frames on program activities and to follow through. For the implementation of an initial 500,000 sq. ft. building I believe that nine months is an appropriate period; however, in a real "live" situation, this period may be reduced or extended depending upon the urgency.

Conclusion

I believe the views offered in this tutorial can serve only as a guide for those who are considering entry into the Shared Telecommunications Systems marketplace. At least a degree of the extensive research and analysis that is vital to the determination of the feasibility of any such project will be saved and that has been my purpose. As each project is different in terms of building size, scope of the services offered and the environment that it is to be established in, no absolute expense formulas can be provided. I cannot therefore include specific costs in the matrices that I suggested in this tutorial.

The reader will also note that I never addressed the costs and other issues relative to the extension of the STS system to the second and third building. Although somewhat different factors have to be considered, the determination of the economic feasibility of serving the second and third building should not be too difficult given that the costs for the initial building system have been well defined. Nor have I addressed the insurance and tax aspects, debt and equity financing, etc. These are subjects that should be addressed by experts in each of the particular fields.

Chapter 18

Office Automation For Shared Tenant Services: Practical Implementation for Various-Sized Organizations

Robert M. Bozeman, Sr.
Wang Laboratories, Inc.

Because it had grown from 50 to 150 employees, a law firm was moving into beautiful new offices in downtown New York City. While the company decision makers were thinking of this move, they were also planning to introduce computers into the company to improve their productivity. With heavy word processing needs and light data processing requirements, they could foresee significant communications problems.

No one in the firm had any great expertise in these areas; so, when they started exploring the issues, they were suddenly faced with trying to match their needs with the wide range of equipment that was available in the market. Compounding the problem were the decisions that would also have to be made about space requirements, flooring and building modifications, electrical and environmental requirements, telephone line and interface needs, and staffing requirements to coordinate and maintain the equipment if it was purchased.

This is a scene familiar to more than a few companies--large and small and across every industry throughout the world.

Confused by what they found in the market, the partners in this law firm decided to move into a "intelligent" building that featured Shared Tenant Services because it offered them a one-stop shopping service for their information technology needs. Prior to Shared Tenant Services (STS), decision makers in companies had only one option: to sort through all of the different telephone and computer packages available themselves. With STS, an alternative now exists: one-stop shopping for all of a company's information technology needs at the site where the company is a tenant.

At the crux of the success of Shared Tenant Services is the reduction of inefficiencies in the installation and use of computer technology. The supplier of Shared Tenant Services expands customer base and achieves greater profitability because of more efficient utilization of equipment. The user of Shared Tenant Services is offered economies of scale enabling even the smallest of businesses to enjoy the advantages of sophisticated telephone and office automation technology.

To address the practical implementation of Office Automation (OA) in a shared tenant services environment, we need to look at two perspectives simultaneously--the supplier's and the tenants'. From there we can develop an approach to implementation that benefits both players.

This approach can best be imagined as a graph with three axes. The first axis represents the type of STS customer; the second, the size of that customer; and the third, the service most likely to be needed. The type of customer can range from a general office setup to one that is voice-oriented, data-oriented, OA-oriented, or both OA- and data-oriented; the size of the company grows from small to large; and the type of services needed can range from simple telephone service to an array of communications and computer services, including word processing and data processing.

By analyzing each of these axes, from the supplier's and the tenants' views, we can draw some generalizations about OA implementation within an STS environment. As with all generalizations, however, caution is always recommended. Doctors may know that certain groups of people are prone to certain diseases, but they still treat patients individually. So too must STS providers.

Tenants' Perspective

The concept of STS puts tenants in a very unique position because it allows them to take advantage of advanced technology without making a heavy capital investment. Services generally fall into three major categories: basic telephony that includes local telephone access, long-distance access, equipment rental or purchase, and voice mail; custom information services that consist of word processing, data processing, electronic mail, and local area networking; and, finally, extended information services that offer such options as local area networking, remote computing services, database system access, wide area networking such as packet switching, and gateways to other teleprocessing services. Within all of these categories, STS also provides tenant-needs consultation and maintenance support with on-site staff.

Some of the obvious benefits of STS, therefore, could include:

--long distance services with least-cost routing and a detailed call reporting program for usage analysis and accountability,

--telephone system and station features for productivity improvements,

--reduced voice and data communications costs with no capital outlays,

--convenient one-stop shopping for communications and computing services,

--24-hour service with on-going, on-site operations,

--elimination of communications management and administrative costs,

--growth capabilities without equipment obsolescence, and

--better use of leased space.

By leasing space in an intelligent building, businesses not only gain economies of scale inherent in sharing resources, thereby saving on costs, but they also have access to efficient, on-site service. These factors significantly affect businesses' productivity.

Is STS For Everyone?

It may not be. Let's imagine our model again and focus on STS customer definition. Customers essentially come in three sizes: small, medium, and large, with the large customers employing over 200 people, the small firms under 100, and the medium-sized companies sandwiched in between.

Each of these sizes also comes in two varieties--single location or multiple locations. All but the largest of companies that have multiple locations are probably good prospects for STS because they show a propensity toward benefiting from the pooling of resources, thus enjoying an economy of scale they could not otherwise achieve on their own. In addition, they generally have little or no technical expertise on staff. Larger firms may have experts at headquarters who have established their own standards and preferences in equipment.

Most STS suppliers, therefore, might look for multiple-location firms as well as for many single-location firms; professional firms; and offices that are phone-driven, that have long-distance volume usage, and that have great upscale communications where the quick turnaround of detailed presentations is critical. These firms might include offices performing predominantly clerical functions; showrooms; professional offices (law, accounting, medical, architectural); manufacturing sites; and service centers that distribute time-sensitive materials.

Financial Considerations

Whether or not a company decides on OA equipment usually becomes a question of money, so let's evaluate some of the financial considerations of STS. The heart of the "smart" building is the digital PBX, which routes calls around an office building and connects outside phone networks. The reason telephone service is included in a discussion about OA implementation is because most of the new PBX's are capable of communicating the digital input and output of computers, thereby carrying voice and data simultaneously over phone lines. This is a key point that will be expanded later in this chapter.

What exactly are the financial ramifications of purchasing a PBX outright? The substantial capital investment and complexities of designing, installing, and maintaining the most effective system suggests that smaller organizations cannot purchase state-of-the-art equipment at the same per telephone cost as larger organizations. Clearly, smaller businesses would benefit from a PBX network, but cannot afford it. STS equalizes the effect by providing that network to many users and allowing the cost to be spread over a wider base.

Another major factor to consider is the operating cost, including both the direct and indirect expenses. The direct costs reflect actual telephone usage. With Shared Tenant Services' buying power, tenants can benefit in the savings due to discounts offered by different carriers, and by the sheer volume produced by the cumulative effect of many users. The indirect costs include the management of the system, such as maintenance and technological upgrades. These, of course, are automatically taken care of in an STS environment.

And, with office rental costs continuing upward, space for such equipment should be factored into a tenant's decision to purchase as well. A PBX for a single tenant may require 100 square feet and cost $3,000 a year for space alone. In addition, special construction may be required for independent air conditioning and equipment security. It has been estimated that for every $1 independent tenants pay for rental space, they would pay about $0.78 for communications.

Assuming a company has 100 employees and each employee averages 200 square feet of space, the minimum area this company would require is 20,000 square feet. According to a recent New York Times report on commercial real estate, rents ranged from $8 to $60 from New York to Los Angeles, with the average leveling out at $30. At $30 per square foot for 20,000 square feet, annual communications costs without STS would be over $450,000. With STS, companies pay for only what they use, and PBX and computer systems equipment is located on the STS provider's premises within the building.

Similar Analogies Can Be Drawn For OA Equipment.

As a theoretical example, basic word processing equipment with standard disk and daisy printer requires about 80 square feet. At the average rental, that factors out to an annual cost of $2,400 on office space alone for equipment. Add the equipment purchase or lease cost, software, maintenance, and internal staff supervision and the cost rises accordingly. With STS, tenants now have an option. They can delay equipment purchase until they know they will get full utilization of their equipment.

Word Processing Options In An STS Environment

In an STS-equipped building, there are a number of options open to tenants who need word processing. First, ignoring the STS capabilities totally, he could come in with his own equipment and hook up his system with the appropriate equipment, cabling and environmental modifications that might be required. Second, using STS capabilities, he could talk with the on-site consultant to determine how best to match his equipment with the STS environment that exists. That may mean using his terminals and hooking them up to the STS computers, or purchasing or leasing equipment and accessing the building's computer. Third, the tenant could simply use the word processing services that would be available through the STS operator.

Only in an STS environment can a word processing service be effective and profitable because of its potential customer base, the PBX data switch, and shared usage of efficient facilities. Easy access and easy use without commitment to major capital equipment that might be underutilized gives these tenants the incentive to use the service on an as-needed basis.

In illustration, at some STS sites, a computer for word processing is configured behind a digital PBX in the building. There are standard asynchronous user terminals attached to digital sets on the subscriber side of the PBX. One-button dialing, port allocation, connection establishment, and detail call record data are managed by the PBX. Any combination of customer-supplied "dumb" asynchronous terminals or PC's emulating "dumb" terminals can be used. Thus, STS customers receive full functions and high performance word processing (WP) on inexpensive terminals at low measured service rates. This allows tenants to use only what they need and insures better utilization of equipment. Regardless of the terminal brand, the word processing files can be consistently formatted, easily interchanged, and editable as a result of sharing standards through the STS PBX.

A user simply dials up a number and enters his or her identity codes. When all operations are complete, the user exits from word processing, which is similar to hanging up the telephone. STS customers are billed based solely on the duration of the WP service call. To print a document, the customer can either use a printer located within his own premises or at the WP Service Center, for which he will be charged for usage only.

Supplier's Side

In Chapter 1, office automation is defined in terms of satisfying the data and word processing needs of customers. As the chapter states: "Most offices today have office automation in the form of data and word processing. However, data and word processing operations often take place independently with little or no integration between the two functions. This results in considerable inefficiency. The preferred office automation system integrates data and word processing operations, making it possible to achieve full data integration capabilities."

That statement can be extended to include all forms of information processing: voice and images as well as text and data. This is an important distinction because that is where the market is driving information processing, and only when all forms are integrated will we achieve the highest level of functionality at the lowest possible costs. That is also why the trends occurring with digital PBX's are critical. They will play a pivotal role in the total integration of all forms of information processing.

Thus, referring back to the three axes of our graph, the third axis focuses on the technology an STS provider can bring to the tenant and that starts with basic telephone needs and extends to advanced telecommunications requirements. From putting a telephone on a desk to installing satellite antennas or microwave dishes on the roof, it's literally a soup-to-nuts operation for the developer and STS provider.

The developer or owner of a commercial site must exercise considerable judgment in providing the necessary infrastructure to accommodate the current and future needs of its tenants. Developers can do that by carefully matching the three axes of our graph. The graph represents the minimum that might be required in any type and size of company. Function is not mutually excluded from these companies, but the suggestion is that because of their size and the type of business they are in they probably would not need or could not afford the technology.

The STS supplier faces several alternatives in servicing an office complex. Suppliers can act as retailers of OA (hardware and packaged software); that is, sell to each user. That, of course, is expensive and not mutually rewarding because there is no strategy for dealing with systems' integration in the facility.

The supplier can introduce turnkey facilities, which implies unusual offers of integrated communications and computer equipment with the necessary skills of implementation supported by deep penetration of key industries such as law firms, where software is highly customized.

Or, the supplier can contract with tenants for on-going services from hardware to software and support. That is the concept of STS.

Services in an STS environment will vary, of course, depending on the provider, but an STS operation suggests that suppliers need:

--a wide range of hardware and software,

--a wide range of skills,

--a central support group to augment local implementation,

--capital investment capability, and

--longevity in business.

The STS environment will vary from building to building, and a variety of experiences and strategies can be referenced to illustrate the implementation of services with an intelligent building.

An STS building can provide three major communications needs. First, there is long-distance telephone access. Some companies who supply STS equipment and services have developed strategic alliances to satisfy these needs.

Second, there are a range of specialty technologies, such as voice messaging, networking, video communications, broadband (LAN) capabilities, and computer systems. And third, there are specialties such as security, environmental controls, and information services. To meet these needs, companies can apply mixed technologies to meet STS needs -- both their own and those of other suppliers of equipment.

As mentioned earlier, the movement toward digital PBX's signals a giant step toward the full integration of information processing needs and is a key element in some equipment suppliers' communications strategies. Traditionally, businesses have used twisted-pair networks for telephone service and coaxial cable networks for access to computers. Applications that require voice, data, or graphic communications that are occasional and switched, rather than practically constant between points, find new PBX capabilities as the best solution.

The great advantage of using a twisted pair wiring network is that it is already installed in many buildings. In addition, it can serve as the basis for a complete voice/data network. Through the simplicity and cost-savings benefit of using a single twisted-pair wiring system of the digital PBX's, users can receive a host of new voice-processing features including least-cost routing, call detail recording, directory/message center services, queuing, camp-on/call back, call forwarding, station hunting, and uniform call distribution.

At present however, digital PBX's will support data transmission speeds of up to only 64,000 bits per second. That's not enough to accommodate data-intensive applications. Relatively constant connections with high-speed throughput, such as required by full-motion video, are best serviced by a local area network, or LANs. Some of the newest LAN's support aggregate speeds of up to 440 megahertz and because they have such bandwidth, they can support full-motion video and imaging capabilities.

Thus, in an STS environment where an LAN company was the supplier, tenants would have the capability of both the local area network for high-speed transmissions and the digital PBX for high levels of connectivity. Some companies will fully support both current and future customers by providing a migration path from the coaxial to the twisted-pair environment and by developing gateways to enable access to either network.

In conjunction with these networking capabilities, some companies also offer an information systems network in their STS buildings that covers word and data processing with the purchase or rental of terminals and/or systems. It also encompasses electronic image transfer, video conferencing, closed circuit TV and the offering of data processing applications locally and sometimes remotely on computers. Other computers with proprietary software may be accessed as well.

Management Of An STS Operation

It's one thing to pre-wire or pre-cable a building for technology, but what happens after the systems are up and running?

The beauty of an STS operation is that its management occurs onsite. Within an STS structure, providers develop facilities for equipment storage and for servicing customers. In addition, an inventory of spare parts is stocked and on-site technical staff are available to insure fast service.

Service centers at each STS site are staffed to work with tenants in the building regarding their communications needs. The centers also might display a wide range of PC's, terminals, and printers, available for rent or sale.

Future

To make the intelligent building become a fact of life within the commercial real estate markets, the trend toward greater interactivity and integration of all forms of information is essential. As a result of greater economies of scale resulting from expanded markets, technology -- even the most advanced -- will be at the reach of even the smallest of businesses. Such achievements will lead to higher levels of productivity and greater economic results for all businesses.

Chapter 19

Planning For Multi-Tenant PBX Installations

Joseph M. Baker
Northern Telecom Inc.

Multitenant service (MTS) is a hot topic among builders and developers. The information in this chapter is intended as a basic guide for analyzing the economic feasibility of MTS.

Multitenant service is defined by three essential criteria --

1. Shared access to "PBX-like" capabilities and value-added services;

2. Shared access to public networks (local and inter-exchange); and

3. Total communications administration which may include the building automation function.

The first two items imply <u>unit cost reduction</u>, thanks to sharing; they provide more functionality to smaller users. Item three implies <u>real added value</u>, the pre-divestiture simplicity of one-stop shopping.

AN MTS ECONOMIC MODEL

MTS system operators may be building owners or developers, system or facilities management companies, or anchor tenants. This model applies to builders, developers and system management firms. If anchor tenants use it, they should evaluate costs in their own organizations before they try to establish pricing scenarios for a "subscriber" base. If an

Exhibit 19-1 Multi-Tenant/Shared Tenant Services
Pro Forma Income Statement

	YEAR ONE			YEAR TWO		
	Local	Long Distance	Total	Local	Long Distance	Total
Revenues						
1 Local Service	XXX	—	XXX			
2 Long Distance	—	XXX	—			
3 Message Center	XXX	—	XXX			
4 Moves & Changes	XXX	—	XXX			
5 Enchanced Services	XXX	—	XXX			
6 TOTAL REVENUES	XXX	XXX	XXX			
Expenses						
7 Equipment Lease Payments	XX	XX	XX			
8 Equipment Maintenance	XX	XX	XX			
9 Use Tax	XX	XX	XX			
10 Cost of Local Service	XX	—	XX	(Continues thru Year 7)		
Cost of Special Access:						
11 - Charge	—	XX	XX			
12 - Surcharge	—	XX	XX			
13 Cost of Long Distance	—	XX	XX			
Salaries:						
14 - Attendants	XX	XX	XX			
15 - Technical Support	XX	XX	XX			
16 - Accounting Support	XX	XX	XX			
17 - Manager	XX	XX	XX			
18 Rent & Utilities	XX	XX	XX			
19 Marketing Expense	XX	XX	XX			
20 Insurance	XX	XX	XX			
21 Other Costs	XX	XX	XX			
22 TOTAL EXPENSES	XXX	XXX	XXX			
23 Earnings Before Tax	XXX	XXX	XXX			
24 Income Tax	XX	XX	XX			
25 Earnings After Tax	XXX	XXX	XXX			

anchor tenant has a large critical mass in place, it becomes easier to develop very competitive pricing arrangements.

The components of this definition can serve as "macro" groupings for specific line items in a "micro" model of MTS operations. Exhibit 19-1 represents the micro model as a pro forma income statement for an MTS operation. No figures are included; only column headings and row line items are given as a framework.

Eventually, customers using this framework may want to furnish their own figures, add or delete line items, and compute their own pro forma profitability projections. Some figures and formulas are given here, however, as examples only to illustrate some of the economic concepts involved.

SETTING A PRICE FOR MTS SERVICE

As with any product or service offered for sale, price is determined by two major factors -- 1) the underlying cost structure and required (desired) return on investment, and 2) "market" price established by competitive offerings.

If considerable competition for the MTS offering already exists, then the price-setting exercise may be simple and pragmatic -- meet your competitor's price (presumably "what the market will bear"). This obviously assumes knowledge of competitive pricing, which may or may not be available depending on the offering.

Centrex Charges

One substitutable, though not equivalent, offering for which price information is easily available is CENTREX service from the local operating company.

Other MTS Offerings

Other substitutable offerings include existing MTS operations located nearby, which may extend service via "remotes" to your building. Pricing for these may be more difficult to obtain.

Stand-alone Key System/PBX Costs

Another benchmark for comparative pricing can be the tenant's cost for a stand-alone key system or hybrid PBX, with telephone-company supplied business lines. Though possibly not as feature-rich as a fully-digital PBX, modern key systems or small hybrid PBX's are highly capable, and may represent the toughest price competition to digital PBX shared tenant offerings. (This is especially true in the early months of MTS operations, when low penetration rates create an uncompetitively high per station allocated cost.)

"BOTTOM UP" PRICING

Even if competing service price benchmarks exist, however, the MTS operator should develop "bottom up" pricing based on the cost of providing service. This methodology involves several steps --

1. Establish the <u>time period</u> for evaluation. The objective for most MTS operations is to break even financially or make a margin of profit, but over how long a period of time? Because of low tenant penetration in the early months of operation (and resulting under-utilization of switch capacity), operations may lose money initially. If initial service pricing is set high enough to fully recover all start-up costs and high per-station allocated costs, the cost may be prohibitive. In the economic evaluation process, examine a range of penetration and switch utilization scenarios. Use an intermediate (3-5 years) timeframe over which to average total start-up and operating costs to arrive at a reasonable price.

An easy way to determine how much time it will take to lease a building is to contact a local chapter of the Building Owners and Managers Association (BOMA). This group maintains current rental statistics for all major U.S. and Canadian markets.

2. Segment the MTS business, in total, into as many logical components as possible (that is, local service, long distance, message center, maintenance/repair, etc.). Try to identify a separate segment for each separately priced service. For example, if maintenance is to be "bundled" into the price of local service as a "no-charge" offering, then no maintenance segment is required.

3. Project a statement of expenses for the MTS operation, covering the selected intermediate time period, stated by year or by month. Segment these estimated expenses, as discussed above, into separate, billable categories.

Some expense items will be easily associated with a given segment (for example, inter-exchange carrier charges are associated only with the "long distance" segment) while others are not associated with any single segment (for example, rent, utilities, marketing expense). These latter expenses must be fully allocated to the respective segments based on an appropriate measure of utilization (that is, square footage, calling minutes, estimated revenues, or a combination of these.)

Once you fully allocate all expenses to segments, you can establish per unit pricing. For example, over the next three years, total expenses to operate the message center, direct and allocated, are projected to equal $423,360. Over this period, approximately 1,512,000 messages are expected to be taken and billed, giving an average unit cost of $0.28. A gross profit margin of 30% is desired for this segment, so a per message price of $0.40 is established.

After you set per unit prices using this method, you can compute total revenue for each segment. Prepare a full income statement, by segment, by year, and compute overall profitability indices (that is, return on sales, internal rate of return, return on investment, etc.).

You can use a personal computer to calculate several alternate scenarios to determine price sensitivity on profitability, and the effects of variable rates of tenant penetration and system utilization.

MTS REVENUES AND EXPENSES

Certain revenues and expenses, when taken together, make up the "PBX-like", or local service, component of the MTS operation.

1. <u>Local Service Revenue</u>: Revenues for providing a standard telephone set, PBX dialtone, access to the local public exchange, repair service, call detail recording, and a standard complement of non-memory intensive features. BOMA statistics indicate that office worker requirements have increased to 224 square feet per person. To estimate the number of lines and terminals required, divide total net square footage by 250. For purposes of the pro <u>forma below</u>, we will assume a price of basic service at around $29 per month per telephone.

2. <u>Message Center</u>: Charge tenants a flat fee per message for an attended message center. Offer unattended messaging systems on the basis of a flat monthly fee. Based upon historic data obtained through answering service bureaus and independent research firms, the following is an example of determining message center revenue:

<u>Message Center Revenue</u>

# of Lines x # of Workdays/yr. x # of Messages/day x
Fee for Message = Message Center Revenue.

1,000 x 252 x 2 x $0.40 = $201,600.

3. <u>Enhanced Services</u>: Revenues for additional "office automation" offerings, such as word processing, data processing, facsimile, photocopying, paging, or video teleconferencing. Consult with appropriate vendors to establish reasonable pricing. Based upon studies conducted by several independent research firms in 1984, approximately 25% of total MTS revenues were generated by such enhanced services.

4. Equipment Lease Payments: This model assumes all major equipment is covered by a monthly lease payment expense. Obtain operating lease rates from any leasing company, then apply them to the total proposed equipment cost to determine this periodic payment amount. Other equipment acquisition options are available, that is, outright cash purchase or a financing lease arrangement. These would have different effects on the pro forma income statement and on the resulting net present value calculations. Consult an accounting or finance expert for evaluation of these alternatives.

Since this equipment supports both local service and network access, this expense item is pro-rated over both components on the basis of relative share of total revenue. In other words, if 30% of total revenue derives from local service, then 30% of this expense will be assigned to the local service segment (column).

5. Equipment Maintenance: Costs for all equipment repair, except any resident technician salaries.

6. Property/Use Tax: These items vary from state to state; check local taxation authorities or tax advisors for methods of estimating tax liabilities.

7. Cost of Local Service: Amount paid to the telephone company for business trunks (CO, DID, private lines). Consult telephone company for specific tariffed rates and regulations.

SHARED ACCESS TO PUBLIC NETWORKS

Three major options exist for an MTS operator to provide inter-exchange tenant service.

1. Switched Access (Figure 19-1)

All inter-exchange calls placed by tenants are routed (by the tenant switch) over "common lines" (CO trunks) to the local central office, where they are switched to the selected inter-exchange carrier (IXC). No private lines are involved. An access charge will apply to each CO trunk.

Fig. 19-1 "Switched" Access

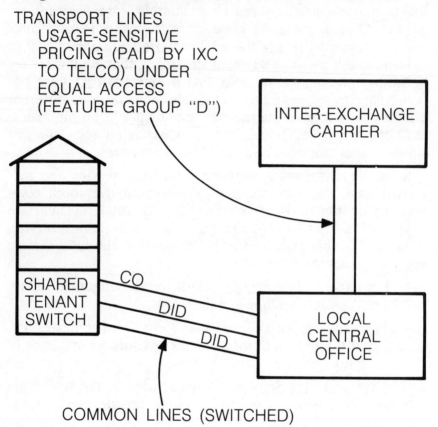

TRANSPORT LINES
USAGE-SENSITIVE
PRICING (PAID BY IXC
TO TELCO) UNDER
EQUAL ACCESS
(FEATURE GROUP "D")

INTER-EXCHANGE
CARRIER

SHARED
TENANT
SWITCH

CO

DID

DID

LOCAL
CENTRAL
OFFICE

COMMON LINES (SWITCHED)

2. Special Access (Figure 19-2)

Sufficient long-distance traffic to justify a <u>private line</u> between the tenant switch and an IXC, may allow "special access."

All long distance (LD) traffic (except overflow) travels over the private line. As presently proposed by the FCC, the MTS operator pays the telephone company for this "special" access in two parts: 1) the going flat-rate for a private line, plus 2) a <u>surcharge</u> to compensate for lost revenues. Our model assumes the MTS operator uses special access.

3. Bypass Access (Figure 19-3)

The whole concept of bypass is one that is sensitive to federal, state and local public utility regulation, tariff restrictions, and in some cases state laws. By this reference, the use of bypass is neither endorsed nor encouraged, but simply noted as an option that has been feasible in some MTS operations. In the example, the connection is via a private microwave link, bypassing all local exchange facilities.

LONG DISTANCE REVENUES

Based upon studies obtained from independent research firms, long distance services normally represent approximately 50% of total MTS revenues. Various pricing techniques are available:

<u>Average Per Minute Rate</u>: A single, average per minute rate for LD usage, regardless of the specific carrier actually selected on each call and regardless of the location called. This rate is based on the average cost per minute plus a built-in margin of profit.

<u>DDD Discount</u>: By discounting off published DDD tariffs, LD pricing is usage <u>and</u> distance sensitive but guaranteed to be lower than direct distance dialing.

<u>Cost of Special Access Charge and Surcharge</u>: The telephone company charges for "special" access lines in two components: charge and surcharge. The charge is the basic costs, paid to the telephone company, on a fixed, monthly basis

Fig. 19-2 "Special" Access

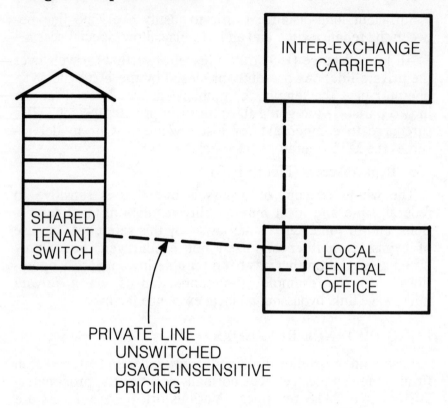

INTER-EXCHANGE
CARRIER

SHARED
TENANT
SWITCH

LOCAL
CENTRAL
OFFICE

PRIVATE LINE
UNSWITCHED
USAGE-INSENSITIVE
PRICING

for the private access lines used between the MTS switch and IXC switch. The surcharge is paid to the telephone company, per private access line. Consult the local telephone company for specific details.

Cost of Long Distance: This is the aggregate of charges from one or more IXCs for network use. The MTS operator is responsible for any discrepancy between the aggregate revenue collected from tenants and the aggregate amount owed the IXCs. A reliable call detail recording (CDR) billing system is essential.

TOTAL COMMUNICATIONS ADMINISTRATION COST

The following expenses, among others, are required to provide total communications administration. No explicit revenue is earned for this added value -- it must be built into the other rates (except for moves and changes).

Moves and Changes Revenues: Moves and changes represent a significant revenue opportunity. Over and above any "free" installation or moves and changes offered, a flat rate may be charged for relocation and reprogramming of station equipment.

Typically, 20% of all telephones terminals are moved in any given year. Most telephone companies charge from $100 to $150 for set relocation, software updates, etc. Because the MTS operators have resident technical support and lower overhead, the same service generally is offered at 50-75% of those rates.

Salaries: Total salary plus fringe benefits for console attendants, resident craftspeople, clerks, managers, and professionals. Use the following guidelines:

Attendants -- Approximately one attendant for every 200 subscribers. Existing MTS operations are currently providing salaries averaging $15,000 plus 30% extra for benefits.

Fig. 19-3 "Bypass" Access

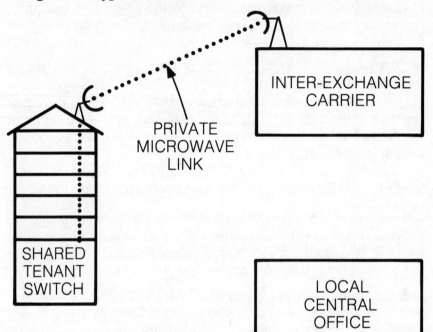

INTER-EXCHANGE
CARRIER

PRIVATE
MICROWAVE
LINK

SHARED
TENANT
SWITCH

LOCAL
CENTRAL
OFFICE

Technical Support -- At least one full-time resident technician is required for the system. An average of $25,000 plus benefits.

Managers -- An average salary of $50-80,000, depending on range of control.

Rent & Utilities: Includes costs incurred for equipment room and office space rental and all related power consumption.

Marketing Expense: All non-salary expenses incurred to attract and retain subscribers, such as advertising, promotional literature, entertainment, giveaways, etc. Expressed as a percent of revenues, heavily front-ended to reflect start-up. Experience shows that for planning purposes, use 12% in year one, 6% in year two, 2.5% in year three and 1% per year thereafter.

Insurance & Legal Expense: Based upon MTS customer feedback, use one percent of equipment costs for insurance estimates. Legal expense will be high during start-up and during the first year of operations.

Other Costs: Your costs will vary according to individual needs. Include a contingency factor of at least 5% of revenues to cover unforeseen expenses. Other start-up costs include numerous items such as organizational costs, licenses, fees, etc.

Income Tax: Because of the important favorable impact which tax loss carryforwards and investment tax credits can have on after-tax cash flow, these factors should be reflected in any economic evaluation. Consult your financial or tax counselor for full guidance.

MODEL DIMENSIONS

By putting this framework on a personal computer spreadsheet, other critical dimensions may be added for greater realism and accuracy. This will also allow an infinite range of "sensitivity analyses" and the development of "breakeven" scenarios for different business assumptions.

TIME

Use at least a seven-year timeframe, with the ability to compute either a <u>net present value</u> of cash inflows/outflows or an <u>internal rate of return</u> for the project. A <u>zero net present value</u> may be used to determine breakeven points for various scenarios (system line size, served square footage, etc.).

PENETRATION/UTILIZATION

How fast and to what extent will a building's total tenant base subscribe to the MTS offering? Recent U.S. statistics indicate that, on average, 65% of tenants in a fully leased project will subscribe to a shared tenant communications system. Run the model several times to evaluate scenarios ranging from worst case (less than 30% participation) to best case (over 80% participation).

Also, what is the initial equipped/wired switch configuration?

Initially over-equipped and under-utilized? Evaluate the economic tradeoff between ordering a fully equipped system up front (to benefit from lower "contract" pricing) and ordering the system modularly, as needed over time (and paying higher, "add on" pricing).

This article was written to provide the reader with a basic framework by which to evaluate entry into the multitenant marketplace. Much time and effort will be required to develop a truly realistic business plan based upon the unique requirements and demands every MTS operator must face.

Chapter 20

Long-Term Legal Concerns About Shared Tenant Services

Victor J. Toth
Attorney At Law

The Department of Justice has recommended the elimination of the MFJ's Line-Of-Business (LOB) restrictions against the Bell Regional Holding Companies (RHCs) entering into information services, manufacturing, other non-telecommunications enterprises and interexchange long distance services. Ironically, it is widely perceived that of the LOB areas in which the RHCs appear to fit most naturally, long distance presents the greatest competitive threat and therefore stands the least chance of being approved. This could be unfortunate. An outright denial of any RHC opportunity in the interexchange market will merely perpetuate the status quo, primarily to AT&T's advantage.

With equal access nearly completed, transmission costs nearing their lowest possible level (subject to further FCC action on end user charges), and yet another round of AT&T price reductions to follow in mid-1987, the competitive interexchange carrier industry does not have much to look forward to.

Moreover, neither state nor federal regulators appear to be softening in their favor. In my opinion, the DOJ recommendation provides an important and much preferred opportunity to influence the conditions under which the RHCs will be permitted to enter the long distance market and, in the process, to rectify the regulatory abuses and equal access problems that have arisen since divestiture. To let this opportunity go by could be a mistake for IXCs and end users alike.

I believe Judge Greene should accept the DOJ's recommendation and let the RHCs enter the interexchange long distance services market. This entry, however, should be subject to the concerns described below.

THE DOJ'S INTEREXCHANGE PROPOSALS

The DOJ recommended that the RHCs be relieved if the MFJ's prohibitions against competing in the interexchange long distance markets under the following conditions:

1. Out-of-Region Long Distance Competition: The RHCs would be permitted to compete immediately for interexchange long distance business in geographic markets outside their respective regions. They would also be permitted to enter long distance joint ventures with other IXCs. However, subject to certain conditions, they would not be permitted to originate or terminate long distance traffic to or from any exchange areas within their regions which are protected by lingering forms of legal, regulatory or tariff barriers to competitive entry or service resale.

2. "Within-Region" Long Distance Competition: The RHCs would also be permitted to offer interexchange services to or from points within their regional markets, provided however, that the market area is free from barriers to intraLATA competition. The DOJ has tentatively proposed the following guidelines for measuring whether an "open" competitive environment will be deemed to exist:

- Subject only to modest certification requirements, the state permits non-BOC carriers, private users and others to provide, either for themselves or for others, all forms of intrastate or interstate transmission and exchange services.

- The state does not require non-BOC carriers to offer universal service.

- The state and its political subdivisions permit BOC competitors reasonable access to state-controlled conduits and rights-of-way.

- BOC services are tariffed and charges are published well in advance, and contract rates for specialty or package services are not offered without public advance notice to other carriers, resellers and end users.

- BOC tariffs discriminate among services only by volume of use, duration of a call, time of day, distance, technical content (e.g., line quality, form of signaling, traffic concentration provided), or other engineering or economic criteria. Charges have to be rationally related to BOC costs.

- BOC tariffs do not discriminate according to who is buying the BOC service, i.e., they are not based on the buyer's identity as an end user, carrier or reseller.

- BOC tariffs do not discriminate according to what is being connected to the BOC line, except as necessary to protect the physical integrity of the BOC network.

- The party ordering and paying for the service (e.g., an IXC or a consultant) need not be the same party that owns the locations to which the service is provided (e.g., customer premises), and the BOC may not insist on dealing directly with the end user. However, the BOC may insist on holding the party it deals with financially responsible for the service.

PRACTICAL EFFECTS OF THE DOJ'S RECOMMENDATION

The DOJ's long distance recommendation is complex and difficult to assess. It purports, for example, to confine the RHCs' near term entry into the long distance resale or networking businesses only to areas outside their regional territories until the DOJ's conditions within their respective regions are satisfied. But this is not the direction that the RHCs are likely to follow. In fact, if anything, the recommendation encourages most RHCs to launch their long distance ventures within, not outside, their region. The DOJ had intended to delay entry into that business, holding it out as an incentive for the BOCs to further open their intraLATA markets to competition and resale.

With respect to out-of-region long distance services, the DOJ intended that each RHC should have maximum freedom to enter this market, subject only to the limitations described above. This would mean, for example, that it could be impractical for Pacific Bell to operate a long distance resale business or to engineer and maintain a proprietary customer network outside California or Nevada. This is because both activities are likely to involve some prohibited terminating calls in Pacific Bell's California service areas. Since Pacific's intrastate tariffs do not permit resale of intraLATA services, and since the California PUC's regulatory policy prohibits intraLATA competition, the DOJ's proposed limitation would preclude Pacific from providing true ubiquitous long distance service even outside California. To satisfy the restriction, Pacific could either handoff its California long distance traffic to another IXC, use customer premise dialers and/or the "10XXX" alternative dialing format to route such traffic directly to another IXC, or it could simply limit its customer proprietary network projects to those which do not include California traffic.

But these possibilities simply impose new restraints and market limitations in what is already an overly competitive and low margin business. Moreover, as US West's "InterFirst" experience recently proved, the RHCs do not have unique strengths and they lack market visibility outside their immediate regions. Thus, they are unlikely to be seriously attracted to the DOJ's out-of-region opportunity, especially so long as the terminating limitation applies.

Another serious misjudgement involves the "within-region" loopholes. Under the DOJ's proposal, six of the seven RHCs (i.e., all but Pacific Bell) would have virtually all the immediate long distance business opportunity that they could reasonably manage right within their own regions. The DOJ's guidelines do not require that an RHC eliminate all competitive and resale restrictions everywhere throughout its region, before offering "within-region" interexchange services. In fact, the guidelines do not even require that competitive restrictions be consistently and universally removed throughout each state. For example, any state regulatory commission or BOC could selectively retain competitive barriers and resale restrictions in certain key LATAs (or possibly, metropolitan markets) while opening others. Under this scenario, the RHCs could establish long distance operations in these "open" markets, while conducting business at usual in the "closed" areas.

Clearly, under the proposed conditions, the RHCs will first enter long distance businesses within their home regions. This will allow them to realize the economies of operating close to home and capitalize on their dominant market presence.

RHC long distance operations within their region, however, pose the the greatest threat of access abuse and cross subsidies -- two problems that were treated too lightly in the DOJ's report. (Essentially the DOJ is proposing to rely on the FCC's new cost allocation guidelines as an adequate safeguard against cross-subsidies, and the DOJ's non-discrimination and cost-based pricing guidelines as assurances against access abuse.)

These were among the vary concerns the DOJ sought to avoid when it devised a scheme it thought would confine RHC long distance activity to modest levels, primarily outside their regions. Instead, the RHCs stand to enjoy as much of both markets as they can reasonably handle for the short term, with little incentive to implement the critical conditions of the DOJ's proposed guidelines. For example, BellSouth could pursue a lot of shared telecommunications service, resale and other proprietary network ventures in selected "open" markets throughout its eight state region before exhausting its long distance business opportunities. Realistically, it could be four or five years before BellSouth would have to eliminate a single barrier or restriction in order to expand its long distance markets within its region.

As a source of further distraction from the DOJ's goal of universal competition, its recommendation also appears to open many interexchange-related business opportunities which fall short of full scale domestic MTS/WATS service-type long distance ventures. These include:

- Maintenance and testing services associated with customer proprietary networks.

- Engineering, construction, operation and servicing of private interexchange point-to-point microwave, cable or fiber projects.

- International long distance.

- Interexchange operator services.

- Provision of selective least cost routing services from a BOC Centrex of other tandem-like switch.

- Joint marketing of interexchange service in connection with CPE or other approved line-of-business activities, or even in concert with another IXC. NYNEX has requested a waiver to be permitted to sell "one stop" long distance services to small and medium-sized telephone users through its computer business equipment stores. There is nothing in the DOJ recommendation that would preclude a BOC from acting as an IXC's agent to promote and sell, under some constraints of course, the long distance services of another IXC for an agency commission.

These activities are currently precluded under the MFJ. They present new market opportunities with none of the headaches of conventional long distance competition. They are also free of any DOJ conditions.

The question is, however, with all these new choices, will the RHCs' plates be so full that they will be indifferent to improving the competitive long distance environment within their regions before entering conventional forms of the long distance business? These ancillary to long distance activities should be held hostage until progress is made toward removing intraLATA restrictions.

WHAT IS THE PUBLIC'S INTEREST?

Strong support for permitting the RHCs into long distance markets exists throughtout Congress, the FCC, NTIA and the Reagan Administration generally. Given the level of intraLATA protection that state regulators have provided the RHCs since divestiture, one can reasonably assume that for all the fields the RHCs are exploring, long distance would be a competitive business venture state regulators could accept. One way or the other, the RHCs will get relief. This is not to say, however, that RHC long distance entry should be supported as a fait accompli, or simply because it is politically fashionable.

The only reasons to support the concept, if they are to be supported at all, concern whether it would be in the public interest and be generally supportive of competition. As to the former, interexchange activity, especially if confined to "within-region" rather than out-of-region markets, is a natural extension of the RHCs' traditional line of business and would substantially restore the one-stop shopping concept for those end users unable or unwilling to adjust to the current multivendor environment. It would also ensure further long distance reductions in all geographic markets beyond what the current access policies along will be able to support. These claims were rejected by Judge Greene four years ago, but their merit is more apparent today.

As for its impact on interexchange carrier competition, RHC entry into long distance should, at the very least, be conditioned on the elimination of still lingering intraLATA and other intrastate competitive barriers and resale restrictions. This should open such market opportunities as greater intraLATA business on a facilities-based or "1-plus" basis. If the DOJ's guidelines are expanded to include, among other things, access protections, the IXCs could also end up with the lowest possible cost of access for themselves. These two developments are unlikely to occur until and unless the RHCs are allowed into the interexchange business.

Further, RHC entry into the long distance business might offer the only viable form of long term interexchange competition to AT&T. The short term prospects for the financial future of MCI and US Sprint are dismal, and the status of the major national resellers is always precarious. It currently makes good sense to create a fallback environment which has potential competitors with the resources, strategic position and technical skills to acquire and successfully operate large, but failing, long distance enterprises.

If something does not happen soon to improve the margins of the nationwide IXC competitors, vast infusions of captial alone (such as IBM might offer to MCI) will be inadequate to guarantee IXC industry survival. In contrast, the RHCs are in position to offer more than the ordinary "White Knight" investor. For starters, they have unique technical and marketing access to a large regional customer base, and more effective billing and collections capabilities. The IXCs or fiber carriers will not be able to acquire what the RHCs have to offer just by going to Wall Street. The long distance community has already seen hundreds of millions of dollars in personal investments, scrapped switching equipment and unpaid telephone bills wasted on an overbuilt resale industry. This unfortunate experience would be dwarfed, by comparison, if the multi-billion dollar investments made by the facilities-based IXCs and fiber operators were similarly forced into abandonment for lack of viable alternative new owners.

FAIR ENTRY TERMS FOR THE RHCs IN LONG DISTANCE

The DOJ's conditions and guidelines provide a good start, but they are inadequate, in some respects impractical and awkwardly drafted. Most seriously, however, I believe they have everyone's interest upside down -- the public's, the IXC's and the BOC's. This is because the DOJ seeks initially to encourage out-of-region long distance activity. The RHCs have no particular strengths in and offer nothing new to this market. Meanwhile, contrary to the DOJ's expectations, the scheme does not create an incentive to open new markets to the IXCs because of the "within region" long distance opportunities.

Therefore, I would suggest the following changes to the DOJ's proposal:

1. The DOJ should require that the RHCs eliminate the competitive entry barriers and service resale restrictions throughout a substantial portion of their total regional operating area -- not just on a LATA-by-LATA basis, before being permitted to offer conventional forms of interexchange long distance service. In determining whether the RHC's region is substantially open to competition, "substantial" could be measured by the number of states in which restrictions have been totally eliminated, or by the percentage of total access lines exposed to unrestricted competition. This modification is intended to substantially eliminate the "within region" loopholes that exist in most RHC regions and which only serve to preoccupy the RHCs in pursuit of long distance niche markets rather than encouraging aggressive elimination of competitive restrictions.

2. Limited long distance-related business activity should be permitted only in state jurisdictions that are totally free of intraLATA competitive barriers and resale restrictions, irrespective of the status of other jurisdictions in the RHC's region. Permissible but limited long distance activities would include, in addition to the long distance-related businesses and the network services described above, any interexchange service that can be provided without the use of either intrastate or interstate switched or special access services, except to the extent such services are necessary to carry RHC's traffic to the POP of an underlying facilities-based IXC, or for incidental amounts of off-network originating or terminating access by individual customers.

3. The RHCs should also be required to resolve the following access related problems:

Access Costs: The RHCs must eliminate all but the Universal Service Fund component (or its intrastate equivalent) of the carrier common line (CCL) charge for interstate switched access and intrastate access charges in any state in which it seeks to offer conventional long distance services. This would reduce to the absolute minimum the cost of switched access service for all IXCs. Once the cost of access for origination and termination is "cut to the bone," the RHC's opportunity to discriminate, cross-subsidize or play other access games with its competitors would be substantially diminished. Also, given the prevailing political and regulatory sentiments against further CCL shifts to end user charges, this might prove to be the only way for the IXC industry to gain further access price concessions.

Access Equality: The RHCs must tariff a rate for switched and special access services for themselves which is higher than the corresponding service charge offered to other IXCs. So long as the RHCs are permitted to offer long distance services directly out of the central office or the access tandem switch, they will always enjoy a technically and functionally superior and possibly simplified form of access. As has been done with AT&T, the value of this access advantage should be quantified and the RHCs required to pay premium rates.

Dual Presubscriptions: The FCC must establish guidelines and practices regulating the method by which the RHCs capture the "1-plus" equal access traffic of their long distance customers. It is now possible for customers to choose multiple primary interexchange carriers (PICs) assigned to each individual line and to have their traffic routed to one or the other as a function of its destination. For example, while it only has been implemented in one state, it is possible to have all "1-plus" domestic calls automatically routed to one IXC and all international calls to another.

One can readily appreciate how an RHC might utilize this "dual PIC" capability to sell itself as the primary carrier in its market for all intra-region traffic (where presumably it would be most competitive), while leaving only the inter-region traffic for the other IXCs. (A reasonable estimate is that 60 to 70 percent of all interexchange traffic is intra-regional.) Considering the potential unfair advantage and opportunity for abuse inherent in such a "dual PIC" environment, at the very least, RHC entry into conventional long distance should await an FCC investigation into this equal access development and the establishment of ground rules for its implementation.

Avoidance of Unnecessary or Excessive Expenditures: To minimize the construction of duplicative facilities and potential excess capacity, to conserve marketing expense and to minimize the opportunity for marketing abuse, the RHCs should be required to consider the purchase lease or other acquisition of existing transmission and switching facilities and the outright purchase of customer base. If the post divestiture era has taught us anything, it has demonstrated that the long distance market is finite and that the long distance telephone business is capital intensive. What is not needed is another round of extravagant outlays of money for facilities construction and a renewal of expensive sales blitzes for market share.

It is also not in the public interest for large facilities-based carriers or smaller regional resellers to fail because of the arrival of yet other new major competitors. To the absolute extent possible, RHC entry into the long distance market should be accomplished in the most efficient and cost-effective manner, and at minimum cost to both the RHCs and their competitors. The RHCs should not only be exempt from any legal antitrust barrier to acquiring, through direct purchase existing facilities and market share, but they should be encouraged to do so.

CONCLUSION

The promptness with which Judge Greene responded to the DOJ's report by establishing a procedural schedule, and the fast track that he placed it on suggest that the Judge was particularly impressed with either the recommendations or the Huber Report upon which they were substantially based. The fact that Judge Greene consolidated the report with AT&T's proposal to shift MFJ line-of-business screening functions from the DOJ to the FCC, further suggests his predisposition to deny most, if not all, of the DOJ's recommendations in favor of a continued case-by-case approach, with the FCC and possibly state regulators substantially involved in the waiver process.

Such an approach would be both wise and reasonable for virtually all line-of-business issues, except RHC entry into long distance markets. It would not be appropriate to defer the threshold long distance competitive question for further consideration by state and/or federal officials. They will have their respective inputs and "the last word," as they are called upon by the RHCs to address and remove the lingering competitive restrictions and to deal with the access issues likely to be imposed as conditions for RHC entry into this market. Moreover, the protracted procedural and legislative processes act to assure the reseller and IXC industry that there will be ample time to compete without the immeidate threat of RHC participation in long distance.

The DOJ's long distance recommendation and the process afforded under the Court's procedures is a more effective, satisfactory and surer process to shape the conditions under which the RHCs are allowed to compete than would be available through legislation -- and legislation is the alternative if the DOJ's recommendation is denied by Judge Greene.

Finally, for resellers and IXCs with an ambition to be acquired, now is probably the time to support the elimination of MFJ restrictions as a precondition for the RHCs to even entertain such thoughts. Realistically, it could take one year following a final decision (and the appeals which will follow) before the RHCs would be in position to accept offers to buy. In short, we are looking at 30 to 36 months down the road -- just about enough time to confirm that a major change from the present industry structure will be desirable.

SECTION VI

Case Studies

Chapter 21

One Financial Place: The Intelligent Office Building From A Multiple Users' Perspective

Kevin Kane
Midwest Stock Exchange

Maximum capabilities at minimum cost in terms of building controls, voice/data communications and office automation: This is the perspective of users of the intelligent office building at One Financial Place.

Located in the heart of Chicago's growing financial district, One Financial Place is unique in that its major tenant, the Midwest Stock Exchange, is not only a user but a provider of shared tenant services to other users in the building.

Another unique feature is that One Financial Place essentially is a "single-industry" building which primarily serves the financial community. Due to the commonality of interests in this field, the building consequently is better able to tailor its services to meet specific user needs.

BUILDING BACKGROUND

Developed by Financial Place Corporation which is a joint venture of the Cassati-Heise Partnership and U.S. Equities, Inc., One Financial Place is a new, 40-story, 1,000,000-sq-ft office building located on Chicago's South LaSalle Street. (See Figure 21-1) Designed to be the city's highest quality office building, the facility includes two public and two private res-

Fig. 21-1 Midwest Stock Exchange

One Financial Place, Chicago, IL, home of the Midwest Stock Exchange, is one of the most intelligent buildings in the midwest.

taurants; a private rooftop social and business club including hotel suites; a sports center with 60-ft swimming pool; a medical and emergency treatment area; and an acre-and-a-quarter landscaped plaza.

The first major building constructed in the financial district since the completion of the Chicago Board of Trade Building in 1930, One Financial Place is the largest building ever constructed on LaSalle Street.

The Midwest Stock Exchange

The Midwest Stock Exchange has provided an efficient, competitive marketplace for buyers and sellers of securities for more than a century. Today, the stock exchange is the fastest-growing and second largest in the country, providing a full network of trading-related services to the securities and financial services industries.

The new trading floor of the Midwest Stock Exchange uses a state-of-the-art, automated switching system to economically and efficiently provide members with integrated voice/data communications and office automation capabilities. Supporting the floor's automated trading and office operations, the switching system serves 3,000 pieces of floor equipment such as telephones, data terminals, and other devices.

All voice and data communications are handled on standard, twisted-pair telephone wiring instead of coaxial cable. The use of twisted-pair greatly reduced the cost of prewiring the trading floor. In addition, equipment can be easily and economically installed or moved simply by plugging it into phone jacks, eliminating the need for running cable to make hardwire connections.

Due to the voice and data-intensive nature of the securities industry, the Midwest Stock Exchange would have needed these comprehensive communicaions and computer capabilities regardless of any other requirements. However, development of the capabilities presented the stock exchange with a business opportunity to market them to other tenants of One Financial Place.

Midwest Stock Exchange therefore combined with One Financial Place's developer and a building automation system supplier, Johnson Controls, in a joint venture to form Financial Place Communications Co. (FPCC). Through this FPCC joint venture, the communications and computer capabilities of the Midwest Stock Exchange are being offered to all building tenants in the form of shared tenant services. These are among the most technically advanced voice/data communications and office automation services offered by any office building in the country.

BUILDING AUTOMATION SYSTEM

The building automation system at One Financial Place provides control of heating, ventilating and air conditioning (HVAC), lighting, and lifesafety and security functions.

HVAC Operations

One Financial Place's HVAC system is cored in its center instead of its corners. This makes possible columnfree floors, resulting in increased space utilization, simplified and easily-maintained wiring configurations, and more economical heating and cooling operations.

HVAC operations are based on the use of perimeter radiant ceiling panels and a variable air volume system on each floor. Used in place of baseboard heating, the energy-efficient radiant ceiling panels provide greater flexibility in office design, increased space utilization through elimination of radiators, and more even, comfortable heating.

Starting and stopping under the control of the building automation system, variable air volume systems on each floor make it possible to substantially reduce electrical costs because they use only half as much power as a centralized fan system. In addition, each floor's independent fan room and variable air volume system enable tenants to heat or cool specific areas or zones after business hours.

Equipped with an oversized, four-cell cooling tower, One Financial Place uses outside air for cooling whenever possible under the control of its building automation system. This decreases the need to operate refrigeration units, resulting in additional decreases in energy costs. Other energy-saving features of the building include bronze-tinted, double-glazed, insulated windows and a granite, sealed, building-exterior envelope designed to withstand outside humidity.

Lighting

Lighting at One Financial Place is designed to provide maximum comfort at minimum cost. The building uses fluorescent fixtures with reflectors and deep-dish, parabolic lenses instead of conventional prismatic lenses. These fixtures provide softer, more even lighting while reducing energy and cooling loads, resulting in a savings in energy costs of 40 percent.

Lights can be turned on and off automatically under the control of the building automation system. However, this control is also subject to the intervention of tenants who, as in the case of heating and cooling operations, can punch in codes on telephone keypads to adjust lighting to meet their specific needs at any time of day or night.

Lifesafety, Security Functions

One Financial Place provides elaborate lifesafety and security functions. Lifesafety is designed around a fire command center which monitors and operates smoke detectors, smoke evacuation equipment, annunciators for voice communications, and concealed sprinklers on each floor. Security includes a sophisticated, closed-circuit, 24-hour, television monitoring system supplemented with electric locks on stair exit doors. In addition, a round-the-clock guard service is provided.

VOICE/DATA COMMUNICATIONS

At One Financial Place, tenants have immediate access to a voice and data communications system that puts sophisticated digital technology to work in their offices. Its efficient features make the most out of every minute they spend on the phone. Its inherent software provides flexibility to accommodate changing communications needs. Its dependability, redundance and use of an uninterruptible power supply provide protection from the hazards of interrupted service. And, it meets standards of security set by the Midwest Stock Exchange.

Voice Communications

Voice communications involve the use of a very sophisticated system to provide tenants with 80 different features such as call forwarding, call hold, automatic redialing, speed dialing, call transfer, direct inward dialing and liquid crystal -- a small screen display that can accept and transmit messages. Providing a partitioning capability, the system furnishes the same security as a privately-owned PBX system. In addition, the system includes detailed call accounting with custom billing reports which enable tenants to better manage communications expenses while ensuring that every billable expense is properly allocated to the correct account.

Another major service provided by the voice communications system is low-cost, long-distance phone calls. The broad purchasing power generated by multiple tenants makes it possible to obtain long distance service at substantially lower costs than would be available to tenants on an individual basis. Contracts are made with various common carriers so that, when a long distance phone call is made, it can be automatically routed through the least expensive common carrier available at the time. In addition, the roof of One Financial Place is zoned for a teleport in which electronic dishes and antennas can be installed for use in satellite communications.

The communications system is also being used to provide a voice mail service. This service permits callers to leave recorded messages at tenant phones. In addition, a phone-answering service to which tenants can forward their calls is being provided. Other services include a secretarial and word processing pool which can be used by tenants to offload work during peak periods.

Data Communications

One Financial Place provides a data communications system which tenants can use to develop a local area network linking computers and other peripherals within their offices. All types of computers equipped with either RS-232-C or RS-449 interfaces can be connected simply by plugging them into jacks provided by the building's universal, twisted-pair wiring.

The multi-functionality of the building's automated switching system makes it possible for different vendors' computers in the same office to communicate with each other. Communications can be conducted with both a mainframe computer system operated by the Midwest Stock Exchange and outside computer facilities.

OFFICE AUTOMATION

Tenants in One Financial Place can increase their office automation capabilities through the use of IBM PC's in various configurations as stand-alone or terminal systems. Using menu-driven, user-friendly programs these PC's provide automated capabilities in areas such as statistical analysis, data base management, electronic spread-sheets, word processing, business graphics, decision support, electronic mail, calendar management and electronic filing. Often these PC's are sold to support various finanical markets and support functions. Customized software services are also provided to meet special tenant needs.

Maximum security is assured for both standard and customized applications through the use of tamperproof access procedures and the segregated configuration of the mainframe computer system which is specifically designed for timesharing operations.

SUPPORT AND TRAINING

Additional support is provided to tenants through a host of productivity-enhancing tools such as facsimile transmission, Telex, high-speed letter printing, graphic plotting, and access to a broad range of financial management and data base systems.

An automatic order entry and monitoring system is used to facilitate quick and accurate responses to tenant requests for equipment, moves, changes and routine repairs. An on-site team of technical professionals gives the assurance of timely emergency repairs in minutes rather than days.

In addition, training is conducted as required in both tenant offices and a classroom facility capable of handling as many as 20 students.

BENEFITS

State-of-the-art voice/data and office automation capabilities are vital in today's fast-changing, competitive, financial industry.

One Financial Place offers a fast, simple and economical way of providing these advanced capabilities to tenants ranging from brokers and dealers to options firms, bank holding companies and the Midwest Stock Exchange.

These capabilities are made available to small as well as large tenants without any requirements for upfront capital investment and at reductions in operating costs varying from 25 to 30 percent.

Furthermore, capabilities can be efficiently and economically expanded in response to growing tenant needs. Normally, tenants move into a building for three, five, ten or even more years. During this time, their needs will usually increase. For example, they may start out with a basic voice communications service and later add data communications and office automation capabilities.

One Financial Place is able to accommodate this growth with current, leading-edge technology. As a result, tenants are relieved of problems that otherwise might occur due to limited capacity or equipment obsolescence.

In addition, One Financial Place is better able to meet user needs, both now and in the future, because its tenants are primarily in the financial industry where requirements are relatively uniform. Consequently, if new capabilities are developed for one tenant, the chances are good that other tenants will also be able to use these same services. Tenant satisfaction therefore is maximized while unfulfilled expectations are reduced to a minimum.

CONCLUSION

One Financial Place is charging one of the highest rentals of any commercial office building in Chicago. Yet the building was leased up very quickly.

Major reasons for this excellent leasing record are that the building represents prime office space and is positioned well in the financial district. However, the voice/data communications and office automation services provided by the building are also a significant factor.

Eighty-five percent of tenants have signed up for one or more of these services. This is a very high penetration rate which indicates that tenants fully recognize the benefits of occupying space in one of the most advanced, intelligent buildings in the country.

Chapter 22

Federal Building East: The Intelligent Office Building From A Government Prospective

Donald F. Miller
Architect

Robert D. Eberle
Administrator

General Services Administration, Northwest Region

Federal Building East in Portland, Oregon, was designed and constructed under the direction of the General Services Administration's Region 10 which is based in Auburn, Washington, and encompasses the States of Washington, Oregon, Idaho, and Alaska. This building is a "high technology" facility and is significant for two reasons. The first and most important is that the building combines all the proven intelligent building technologies with exciting flexibility. This design provides totally integrated building controls through the PBX voice and data communications system including an environmental management control system, a fire and life safety control system, and lighting controls. The other reason is that Federal Building East clearly demonstrates the important role that the Federal Government's General Services Administration is playing in bringing new ideas and technology into use for the benefit and enrichment of all the taxpayers of the United States.

PROJECT TEAM LAID GROUNDWORK

Months before architects submitted design recommendations for Federal Building East, a General Services Administration Region 10 high technology task force was formed consisting of an architect-project manager, structural, mechanical, electrical, and telecommunication engineers and space management specialists.

The responsibility of the team was to lay the groundwork for the project by developing a "Design Building Program." This involved preparation of a single document covering codes, utilities, regulations, services, special design considerations, tenant requirements and budgeting, i.e., everything involved in final design and construction.

BUILDING AUTOMATION SYSTEM BECOMES STAND-ALONE

Initially, it was not anticipated that Federal Building East would be a "high-tech" project. The building's computer capabilities were originally planned to be provided through an on-line connection with an older computer system installed at theGreen/Wyatt Federal Building in downtown Portland.

However, after the "Design Building Program" was completed, the decision was made to install a stand-alone, computerized building automation system in Federal Building East. This decision made it possible to develop a truly state-of-the-art building. Based on this decision, a number of design program changes were made to upgrade the building systems while reducing long term operation costs. The task force worked out the details and design considerations of energy use, lighting, heating, air conditioning, communications, background noise levels, safety, security and other factors.

As a result of this preparation improved communications were achieved between the General Services Administration, the prime tenant, the contract architectural firm, and other key players involved in the project. This had a significant impact in getting Federal Building East on track quickly and in the right direction in terms of a longer useful life, lower life cycle costs, energy conservation, leading-edge communications, efficient use of space, and higher levels of worker productivity.

BUILDING AUTOMATION SYSTEM SERVES AS HEART

The Building Automation System (BAS) is the heart of operations. The BAS is a central processing unit whose primary job is to obtain maximum efficiency and life from all of the components in Federal Building East's heating, ventilation and air conditioning system. The BAS continually monitors component status through a system of remote field panels located on each floor. The remote field panels are in communication with the BAS and directly connected to components in the HVAC system, including fans, chillers, pumps and air handlers. The field panels provide redundance in the system with full operation in case the central processing unit should go off line for any reason.

Using a network of sensors, the BAS scans areas throughout the building and reacts accordingly. For example, if the system detects a buildup of carbon monoxide in the parking garage it automatically turns on exhaust fans.

BAS ALSO MONITORS EQUIPMENT CONDITION

In addition to monitoring component status and controlling operations, the BAS monitors equipment condition, checking for overheating or excessive vibration and alerting maintenance personnel when a part needs service or preventive maintenance should be performed. This helps to lower maintenance costs tremendously since it is now possible to perform preventive maintenance based on manufacturer's recommended hourly use schedules rather than on a calendar schedule, and problems can be checked and corrected prior to a major failure.

HEAT RECLAMATION IS ANOTHER BAS TASK

Waste heat reclamation is another major task of the BAS, because GSA heats Federal Building East using reclaimed heat from computer equipment, lights, people and solar gain. Fossil fuel is used to heat the building during its start-up hours of operation when the heating plant in the adjacent building is tied into the new building's system to bring it up to a comfortable level before tenant occupancy. At this point, the computer shuts off that heating system and Federal Building East operates on its own unless severe weather conditions require assistance.

SOFTWARE IS USER-PROGRAMMABLE

The software used by the BAS is user-programmable thus enabling the building operator to fine-tune efficiency levels as technicians become more familiar with the building's unique character.

LIGHTS ARE ALSO MANAGED

Careful management of lighting is an important contributor to achieving the goal of reducing kilowatt consumption at Federal Building East to a fraction of that normally used in a building of similar size. Motion detectors in offices turn off lights if they don't sense movement for five minutes. On weekends, employees who work in open office areas dial the system from their telephones, enter a code from the telephone keypad, and turn on lights in their immediate areas instead of illuminating half a floor or more. Furthermore, the system is programmed to check regularly and turn off lights in unoccupied areas.

TENANT FLEXIBILITY

The prime tenant estimates that 30 percent of its employees typically move each year. In the past, this contributed to significant losses in worker productivity which were estimated to total as high as 600 work days annually. Repositioning desks, walls, adjusting air conditioning, restringing cables, dealing with limitations of permanent electrical conduits, sorting out phone numbers . . . all of these contributed to the loss in efficiency.

Responding to these problems, GSA developed a solution involving both software and hardware. Most of Federal Building East, housing 2100 people in 7 stories of open office space, has raised access floor which is a major aspect of the building flexibility. Beneath the access floor is an open space which is used for all power communication and data transmission lines.

Open office areas are divided into bays approximately 30 x 32 feet. Typically, each bay accommodates eight work stations. These bay layouts are stored in a Computer-Aided-Design (CAD) system so that space planners can see the effect of proposed changes before they are made and fine tune rearrangements. Moves are accomplished in a fraction of the time previously required because of the unique floor system that is used throughout the Federal Building East.

Two service cores carry power, communication and data lines between floors. On each floor, lines branch out from the cores and run directly to individual work bays. Power then runs to each work station in series. Communication cables also extend to each bay from the distribution core with individual runs to each work station. The floor above this access area is a grid composed of individual panels that can be removed easily to allow access to the wiring underneath. Individual panels have a cable pull-through which may be oriented in any direction, simply by rotating the panel. It is estimated that this unique system of cabling and distribution is reducing the cost of worker moves and changes by as much as 75 percent annually.

Above each worker bay, access space above the visible ceiling accommodates ventilation ducts, fire sprinkler plumbing and lighting. Modular light panels are used throughout the building and can be relocated, added or removed to meet specific needs. Work station furniture provides well-defined sound isolation areas for employees. White sound is added to improve ambient noise dampening and provide a sense of privacy, free of distractions.

SAFETY, SECURITY IS VITAL

Safety and security of employees, equipment and records is a vital job at Federal Building East. Exterior access points throughout the building are monitored by electronic sensors. Interior movement is monitored by closed circuit TV. In addition to a lobby monitoring station which is manned 24 hours a day, security is monitored around the clock by personnel at Portland's Green-Wyatt Federal Building downtown via microwave.

FIRE CONTROL IS EXTENSIVE

The first defense against fire is an extensive smoke detection system with sensors located in the access ceiling and floors. This is backed up by pull alarms and heat-sensitive overhead sprinklers. In case of a fire, the BAS automatically notifies security personnel and the fire department while returning elevators to the lobby, closing fire doors, unlocking building exits, and broadcasting prerecorded messages to direct occupants to smokefree areas or escape routes.

The BAS also takes control of the HVAC system to counteract the spread of smoke. Exhaust dampers in the fire zone are opened and supply dampers are closed. In adjacent areas, supply dampers are opened and exhaust dampers are closed. Pressurization fans are activated automatically to positively pressurize stairwells, elevator shafts and other areas around the fire, keeping smoke from entering these spaces and allowing people to move to safety.

Information about the source and spread of the fire meanwhile is fed back to the command center where it is evaluated and used by firefighters in controlling the fire and directing the safe evacuation of personnel. A remote annunciator/command terminal is also available to firefighters. This terminal, located in the fire control room, gives firefighters manual control over a variety of fire, security and HVAC points.

STATE-OF-THE-ART COMMUNICATIONS ARE FEATURED

In this age of communication, the ability of people to communicate with each other inside a building and with others outside, or even halfway around the world, is of utmost importance. Federal Building East is completely state-of-the-art in this crucial area. It has integrated voice/data capabilities, analog and digital microwave transmission, data communications, office automation, and the ability to transmit data at rates up to 56,000 bits per second. The cost of data transmission is reduced greatly through the shared use of modems located in the PBX room. All voice stations have the ability to transmit voice and data communications simultaneously over the same two-pair of wires connected to each telephone.

The telecommunications system is a sophisticated fourth-generation electronic switch. Capable of handling 5,000 users, the system provides enough capacity to service every federal worker on the East side of Portland. Voice mail, access to electronic mail, gateways to local area networks, access to the BAS and an automated directory are also made possible by the system.

FEDERAL BUILDING EAST IS HIGH-TECH BENCHMARK

Federal Building East is making history within the General Services Administration in particular and the Federal Government in general. It is a benchmark by which all future Federal construction projects will be measured. It is an intelligent building, a high-tech building, a prototype for the future whose success will be measured in reduced energy costs, improved worker efficiency, better productivity and lower life cycle costs.

Chapter 23

California Center:
The Intelligent Office Building From
A Developer's Perspective

Charles C. Saunders
Norland Properties

From a developer's perspective, the intelligent office building is a way of getting a competitive edge in a soft office building market.

This is the case throughout the country but particularly in San Francisco where a moratorium has been imposed on high-rise construction. Many buildings were constructed in anticipation of this legislation, resulting in a glut in high-rise office space.

Consequently, it is very important today to provide an office building with additional capabilities that will increase its appeal and distinguish it from other buildings in the marketplace. Tenants realize that some of the most expensive office costs are personnel and utilities. Any opportunity to reduce staff and increase productivity by using building-furnished automated and hi-tech services will receive serious attention.

The intelligent office building uses many sophisticated building automation systems that provide tenant luxuries such as maximum comfort conditions and HVAC operating efficiency at minimum cost, building life/safety, lighting and energy controls, and security functions.

The intelligent office building also provides tenants with a broad range of state-of-the-art voice and data communications services. Such services include low-cost, long-distance voice communications, access to local voice and data networks, and high-speed data transmission capabilities. In addition, a wide variety of office automation services including word processing, voice mail, message center, local area network and other capabilities are offered. Having a shared tenant services system in place, therefore the provision of a "switch" for the building, enables tenants to achieve higher space efficiency and thus rent savings through the elimination of the need to install their own equipment.

All of these high-quality services are provided to tenants from a reliable single source, permitting them to operate in an economical, efficient, one-stop shopping mode. A major portion of the normal initial capital outlay for new equipment by the tenant is removed and operating costs are lower due to the bulk buying leverage of the developer. Furthermore, continual on-site service and support is readily available and capacity is provided for smooth, cost-effective future growth.

It is attractive new services such as these that enable an intelligent office building to stand out from the crowd in the eyes of prospective tenants. A case in point is Norland's new intelligent office building in San Francisco called California Center.

BUILDING BACKGROUND

California Center is a development of San Francisco-based Norland Properties. Founded in 1982, Norland Properties is a member of BKS group of companies, a worldwide organization with investments in real estate, oil, gas and high technology companies. Norland is a subsidiary of BKS Realty, which manages the group's U.S. investments.

Located at 345 California Street, California Center is San Francisco's last, major, mixed-use high-rise structure. Rising from the heart of the West Coast's most prestigious business center where today's leading companies of Europe, the Americas and the Asia/Pacific Basin maintain major presences, the building features a dramatic hexagonal design and exquisite attention to detail which make it a beautiful and time-less addition to San Francisco's skyline.

Rising 48 stories above two subterranean parking levels, California Center consists at ground level of a four-story podium base containing 7,700 sq. ft. retail space, The Mandarin Hotel lobby and bar and reception area; banqueting and restaurants on the second floor and two floors of commercial space on floors three and four. Fifth through thirty-fifth floors are composed entirely of commercial office tower with floor sizes of 17,500 gross square feet for a total premium office space of 580,000 square feet. The next two floors are mechanical and hotel service areas. Capping the building are two 11-story towers reaching from 38th to 48th floors. Connected by a unique glass skybridge, these towers house a 160-room grand luxe hotel operated by The Mandarin Oriental Hotel Group, headquartered in Hong Kong.

Specific capabilities of California Center were developed with more in mind than just the requirements of Norland Properties. Norland had entered into an unusual partnership with the San Francisco law firm of Morrison and Foerster under which the firm became a part owner of its floors in the building as well as its anchor tenant. Consequently, it was of primary importance to include this firm in the decision-making process to insure that its needs were met.

Both Morrison and Foerster and Mandarin Oriental Hotels had expressed an interest in the development of telecommunications, data communications and office automation capabilities as an integral part of California Center. Norland saw this as an opportunity to meet their needs while developing services that could be offered to all tenants.

Coordinating efforts with Morrison and Foerster, the hotel, and building architects, Norland worked on the development of state-of-the-art building automation, life safety, lighting, security, voice/data communications and office automation systems. Based on these successful efforts, California Center today is one of the most technically advanced, intelligent operational buildings in San Francisco.

BUILDING AUTOMATION SYSTEMS

California Center operates with two Building Automation systems (BAS). The first system provides tenants with maximum control of HVAC, lighting, and energy conservation equipment at the lowest possible cost. The other BAS controls the various building life/safety systems in order to meet or exceed San Francisco fire code requirements which are among the strictest in the nation.

HVAC Operations

California Center's HVAC equipment includes two main heat exchangers which convert steam supplied by the city into hot water for heating the building; a central refrigeration plant utilizing two large, electric-driven water chilling units; and a supplemental 215-ton cooling tower which is exclusively used to cool computer rooms and areas requiring overtime air conditioning.

The building's air handling system consists of five variable air volume units, each of which is operated under the control of a digital system controller connected to one of the building automation systems. A sixth digital controller is used to control operation of the chilling units and cooling tower.

Constantly self-regulating, monitored and controlled, variable air volume units provide the best possible comfort at the least possible cost. Tenants are provided with individually-zoned heating and air-conditioning controls through the use of variable air volume boxes serving six interior and ten perimeter zones on each commercial tenant floor. Conditioned interior air is supplied through slots in fluorescent lighting fixtures while perimeter air distribution is through linear slot diffusers located at window lines. Return air meanwhile is brought back through heat extract-type light fixtures.

Several energy management programs are also used to further improve operations and reduce costs.   One program provides a "warm-up" mode for HVAC system operations. Based on outside temperature, this program starts heating or cooling equipment as late as possible in the morning while insuring that the building attains proper comfort levels prior to occupancy.

Another program monitors the temperature and humidity of outside air and mixes it with return air as necessary to achieve the most economical mixed-air source.   A supply air reset program is also used to monitor heating and cooling loads and achieve increased operating efficiency while maintaining desired comfort conditions.   In addition, electrical consumption is metered by floor, making it possible for tenants occupying whole floors to obtain displays and print-outs of their power usage.

BAS

Incorporating all of these programs and monitoring/control points into a central Building Automation System simplifies the building engineers job and also saves the owner money by reducing the energy costs and manhours needed to monitor the complex systems that make up today's modern high rise office building.   The dynamic color graphics of the BAS provide a quick reference to alarms and conditions of fans, boilers, chillers and other mechanical systems.   It is almost an idiot-proof training tool that allows new building engineers to become familiar with the different building systems.   Like the old adage says, "A picture is worth a thousand words."

Life/Safety System

Meeting or exceeding all requirements, the California Center life/safety system, located in the Fireman's Control Room on the ground floor, monitors close to 500 fire detectors.   Each of the building's 48 floors is equipped with a minimum of six fire detectors while another 200 are installed in hotel rooms and mechanical areas.

The life/safety system also monitors four pull stations located on each floor and flow and tamper switches installed on the building's automatic sprinkler system.   The fire communications system provides four fireman's phone jacks per floor and audio to as many as 34 speakers on each floor, based on tenant partition layout.

If a detector is triggered or switch activated, the system automatically sounds an alarm on the affected floor while notifying the Building Engineers office, the Security Console, and outside agencies of the alarm condition.   In case of fire, the system automatically operates in a shared mode with HVAC equipment to open all exhaust dampers on the fire floor while closing them on floors above and below.

This damper action evacuates smoke from the affected floor as quickly as possible while pressurizing the others to prevent the smoke from spreading.   Indicated with illuminated exit signs, all stairwells are pressurized to provide a safe avenue of escape.   In addition, all elevators are automatically recalled to the first floor and fire-rated doors swing closed by cutting off power to the magnetic door holds which normally keep them open.

Lighting Control

Typical floors at California Center are equipped with 2' x 4', three-lamp fluorescent lighting fixtures.   Upon request, building engineers can centrally control lighting in perimeters and core zones at any one of four different levels ranging from "off" to one, two or three lamps "on."   This makes it possible for a full floor tenant to achieve significant savings in utility costs by requesting all or partial lights off at specific times. Further cost savings are provided through the use of low-energy ballasts and energy efficient flourescent tubes in all lighting fixtures.

Security Systems

The building entrance doors and elevator cabs have an after-hours card key access. This security system can provide access on 20 status levels in eight daily time zones to a minimum of 2,000 separate identification card numbers. Each card can be used to unlock a door or activate an elevator button only if the identification number on the card is current and valid, the status level of the cardholder is correct, and cardholder is entering or existing within a correct time zone. In addition, a printer tracks the access point, time of day and day of week. System memory capabilities are used to record how many times each card is used in a given period and if unauthorized entry was attempted. An emergency power source activates the memory and time clock used in the system for up to eight hours in case of power failure.

All stairwell doors and street level exits are monitored. When any stairwell or exit door is opened, an alarm is activated in a guard room and a visual indication is displayed on a video monitor. The monitor indicates the particular door which has been opened, the floor on which it is located, and its present condition. Once in the stairwell, in non-emergency situations, the only exit is at street level through a series of additional alarmed doors.

Located in subterranean levels, the building's parking garage is operated on an attendant-park basis. The presence of parking attendants and security cameras meets most security needs during normal business hours. After hours, when only one attendant may be on duty, steel roll-down gates are used to close the garage. Access is then provided by an intercom and an electronic release of gates by the attendant. A detection device is recessed into the garage driveway to alert the attendant to vehicles entering or leaving the garage during slower traffic periods.

A shuttle elevator operating between parking levels and the commercial lobby on the ground floor is card-accessed during non-business hours.   Fire stairwells that lead from lower to upper levels are electronically monitored and security cameras are used to survey the lower level areas where cars and the public enter and leave the building.

The building's loading dock and trash storage areas are monitored directly by the adjacent security office.   Manned 24 hours a day, this security office is the focal point of all video and electronic monitoring operations as well as all fire alarms and emergency signals.

VOICE/DATA COMMUNICATIONS

California Center provides tenants with comprehensive, state-of-the-art voice/data communications services from a single source.   Furnished through an integrated voice/data switch, services include local and long distance service, high-speed transmission and on-site maintenance and administration.   Addressing virtually every communications need of any building tenant, these capabilities result in considerable cost savings to tenants because they are provided as economical shared tenant services.

Voice communications are based on the use of a discount long distance service which is capable of providing high-quality transmission and detailed usage reports while eliminating access code requirements and monthly minimum usage charges. All long distance calls can be recorded on reports detailing the date, time, number called, duration and cost.   This makes it easy for tenants to both control charges and either apply them to in-house projects or pass them on to clients for whom work has been performed.   In addition, a demand billing capability can be provided which permits tenants to obtain an up-to-date phone bill for review at any time.

Provided through the same voice/data switch, high-speed data communications can be economically and efficiently conducted between pieces of equipment in a tenant's office or with equipment located anywhere else in the country through a modem pooling arrangement.   Access to external data bases in the legal or in any other field can also be arranged. In addition, a local area network capability which facilitates communications between equipment furnished by multiple vendors is available for use by tenants.

OFFICE AUTOMATION

A number of office automation services are also offered to tenants of California Center. One service is a word processing program which includes an integrated forms capability that makes it possible to call a prestored form from memory, enter data in the form on the screen of a CRT terminal, and then either print the form or transmit it to another location.

Another optional service is voice mail, which is a sophisticated electronic message sending and receiving system programmed into telephones. Voice mail enables tenants to receive and record messages whenever they are out of the office or otherwise unable to answer telephones. Stored messages can be retrieved 24 hours a day, seven days a week, from the tenant's office, home or any other location where there is a conventional pushbutton phone. Voice mail can also be used as an advanced message distribution center for sending voice memos to as many as 125 other system users by simply dialing a two-digit number.

A message center is an additional service provided as part of its office automation capabilities by California Center. Using advanced computer technology to store detailed instructions on how telephones should be answered, specially-trained attendants provide personalized, intelligent handling of all tenant calls.

HOTEL SYSTEMS

The Mandarin Oriental Hotel atop the California Center has its own HVAC system. This equipment includes thermostatically-controlled, four-pipe, fan-coil units installed in each hotel room and two roof fans which are respectively used to pump fresh air into rooms and tempered air into corridors.

The hotel operates its own specially-configured, integrated voice/data switch. This switch is used to serve three telephones in each room. Operating off one line, like extensions in an office or home, these phones include a unit by the bed and a wallphone in the bathroom as well as a writing desk phone.

The switch also has the capability to handle data communications, making it possible for selected rooms to be equipped with personal computers or integrated, CRT workstations which hotel patrons can use to access data bases or communicate with their offices. In addition, the switch creates the possibility of establishing an executive service area or business center with advanced data processing and communications capabilities. Voice mail and message center services are other automated functions which can be provided by the hotel.

CONCLUSION

An intelligent office building such as California Center has higher front-end costs from a developer's standpoint. But it provides tenants with economical high-tech services which can't be matched by conventional office buildings.

Enabling tenants to increase personnel productivity while saving money, these services make it possible to lease intelligent buildings quicker. This is very important to a developer because it can result in considerable savings at today's high interest rates.

California Center, for example, was already 50 percent leased when it was occupied in February, 1986, based on letters of intent and actual signed leases. Certainly, the prestigious business center location of the building and its high-quality amenities were important. But the intelligent services provided by the building were also a key factor in achieving this lease rate.

Furthermore, in addition to speeding-up leasing, the high-tech services provided by an intelligent building will be a source of continuing revenue for a developer. This revenue eventually offsets front-end costs, resulting in a payback on the intelligent building investment.

Based on our experience as the developer of California Center, we believe it will be almost impossible to construct a non-intelligent building in the future. Perhaps there will still be a place for non-intelligent buildings in the "boonies." But, in the urban marketplace, intelligent buildings will be the only economically feasible way for developers to go.

Chapter 24

Plaza Towers:
The Intelligent Office Complex In A
Suburban Environment

Michael Schulman
Otis Development Company

The first truly intelligent building in a suburban environment is now being developed in Schaumburg, Illinois.

Called Plaza Towers, the intelligent building is designed with a high profile as an architecturally significant structure. In addition, the building provides a high level of amenities and functionality at a relatively low occupancy cost.

Plaza Towers operates with five stand-alone, interactive systems. These are a building automation system providing energy management; a security system; a fire management system; a telecommunications system; and an office automation system.

Each of these is a stand-alone system capable of operating independently of the others. Consequently, if one of the systems fails for any reason, the other systems will continue to function, providing tenants with a continuum of service.

However, although they are stand-alone, the systems are also capable of interactively communicating with each other because they were simultaneously designed and developed on an integrated basis.

This interactive communications capability makes it possible to achieve more economical and efficient system operations because duplication of wiring, control devices and other types of equipment is minimized if not eliminated.

Interactive communications also improve control capabilities while making control operations easier and more effective through the use of simplified logic.

Most importantly, system integration will permit enhancements to be easily and efficiently made as new technological developments become available in the future. This insures that the building will be a living, growing entity which will be able to constantly improve its capabilities to better meet tenant needs.

Overall, Plaza Towers is an intelligent building which is totally tenant-driven in providing immediate fulfillment of the requirements of its medium-size and large corporate tenants on an as-needed basis. The building is dedicated to providing these tenants everything they need in terms of comfort, lighting, security, fire management, telecommunications and office automation capabilities at not only the lowest possible cost but also without the need for any upfront capital investment.

SUBURBAN ENVIRONMENT

Plaza Towers is located on 18 acres in the heart of the dynamic Woodfield area of the Chicago suburb of Schaumburg. The fastest-growing suburban area in Illinois, Schaumburg will be the second largest city in the state in population by the year 2000.

A major site of headquarters as well as regional offices of Fortune 500 and 1000 companies, Schaumburg is also growing rapidly in commercial office space. Adding an average of 1.2 million square feet a year over the last four years, the suburb currently has 9 million square feet of commercial office space in buildings ranging from three to 20 floors.

An easy 15-minute drive from O'Hare International Airport, Schaumburg is easily accessible from throughout the entire Chicago area by means of expressways and other major highways as well as public transportation via commuter trains and buses. Containing nine hotels offering more than 3,000 guest rooms, the suburb is also the site of Woodfield Mall, the largest enclosed mall in the country with 2.4 million square feet of fine shopping facilities.

More like a downtown area than a typical suburban environment, Schaumburg has a unique and progressive city government that has intelligently planned for growth in terms of office building developments such as Plaza Towers.

PLAZA TOWERS OVERVIEW

An architecturally significant building which is equal to or better than any other suburban Class A building in the country, Plaza Towers features two, blue-cast, reflective glass towers that rise 285 feet above the site which already is the highest point of ground in the Woodfield area of Schaumburg. (See Figure 24-1) A band of natural-toned, precast concrete wraps around the base of the building, providing a warm contrast to the dazzling, cool towers above.

The interior environment at Plaza Towers is equally exceptional. All trim is gleaming stainless steel. All office doors are full height while entry doors are supplied with mortised locks. Thinline horizontal blinds cover each window and there is lavish use of marble and mirrors in lavatories.

Containing about 20,000 square feet of office space per floor and 850,000 square feet in total to serve up to 3,000 people, the twin towers are offset for privacy and connected by a four-level, skylit atrium. Filled with tropical plants, shops, restaurants and other facilities, the atrium brings natural light below grade to a cafeteria on its lowest level.

Eight, high-speed, computer-controlled elevators in each tower operate on a demand-sensitive basis so as to provide the shortest response time possible. Two five-level, covered parking structures which are connected to both towers and each other contain 957 spaces each. Surface parking provides for an additional 494 cars.

Outside Plaza Towers, the setting is as dramatic as the building itself. A full 40 percent of the land around the complex is

Fig. 24-1 Plaza Towers

An Artist's rendition of Plaza Towers, Schaumburg, IL, which will feature intelligent building amenities.

devoted to green areas for the exclusive use of tenants. There are more than 7 1/2 acres of open landscaping, including a lake surrounded by a lush lawn and tree-shaded patio areas. In addition, jogging paths and nature trails rim the property.

BUILDING AUTOMATION SYSTEM

Plaza Towers has a building automation system (BAS) which provides energy management capabilities as well as direct digital control of heating, ventilating and air conditioning (HVAC) equipment. In addition, the BAS is used to monitor and control all security operations.

Energy Management

Plaza Towers is equipped with a low-temperature, variable air volume system which uses multiple vaneaxial fans to modulate air flow in accommodating changing cooling and ventilation needs. Air handling fan rooms are located in the top and bottom of each tower and either room can service any floor, assuring a continuous source of ventilation at times when maintenance is performed.

Heat is provided around the perimeter of the building through ceiling-mounted, fan-powered variable air volume boxes which include electric reheat coil control and direct overhead air distribution.

Cooling is provided by a combination of centrifugal, water-cooled chillers and an innovative, highly energy-efficient, ice storage system. At night when electrical rates are lowest, the chillers supply a low-temperature, glycol-water mixture to the internal piping system of each of 18 closed ice storage tanks. Ice is formed on the surface of the piping and is allowed to build to a significant thickness. When sufficient ice is formed, the system shuts down to await daytime demand for cooling.

During the day, the glycol-water mixture is circulated between air handling system coils and ice storage tanks, causing the ice to melt and produce cool air. Colder and drier than the air produced by a conventional system, the cool air provides increased comfort while making it possible to use smaller fans and ducts.

The chillers are also used to provide added cooling capacity on days when the ice storage system is not able to satisfy the total air conditioning load. However, the chillers are still smaller than would otherwise be required by a conventional cooling system and therefore help to further reduce consumption of electrical energy. The end result is substantial energy savings and reduced occupancy costs.

All critical operating data related to the ice storage system, such as ice thickness and temperature, are monitored and recorded by the BAS to insure maximum efficiency and performance. The BAS also provides optimal start/stop capabilities for HVAC equipment of which the ice storage system is an integral part. In addition, the BAS monitors space and system temperatures, records total building energy usage, and produces trend graphs.

Another intelligent building capability provided by the BAS is individual energy metering by tenant. Energy used by tenants who operate after-hours, for example, can be monitored by floor. This makes it possible to charge back added facility costs to these tenants while enabling other tenants to pay less.

Integrated with the telecommunications system, the BAS also makes it possible to use telephones as input/output devices to monitor and control energy and lighting functions. A tenant who is staying after-hours, for instance, can pick up the phone and "dial" for continuing heating or cooling by floor or even by zone based on space design and tenant needs.

The same phone capability can be made available for the operation of energy-efficient, three-lamp, parabolic lighting fixtures on a floor or zone basis. Ultimately, it will be possible to integrate personnel sensors with phones so that activation of energy and lighting functions will be totally automated.

Integration with the telecommunications system also enables the BAS to automatically page essential building staff members on-premises via pocket pagers or call them offsite with programmed messages in case of a critical mechanical problem. In addition, the BAS is used to monitor critical functions and environmental conditions of the telecommunications system to help insure continuous, problem-free service.

Security

The computerized security system provides security in the form of after-hours access control based on the use of card readers at entrance doors and even tenant facilities if desired.

The system also point-monitors all outside entrances and other critical areas such as the main equipment room.

Closed-circuit television monitoring is provided for loading docks and other public areas.

Security personnel are automatically paged or called, and additional calls are placed off-site to agencies such as the local police department in case of a security problem.

Full reporting capabilities in the form of screen displays and printed reports are also provided by the security system. Showing complete entry and exit detail in terms of personnel involved, locations and times, reports can be tailored to meet individual tenant needs.

FIRE MANAGEMENT SYSTEM

Required to operate as a separate, stand-alone system by city ordinance, the fire management system provides state-of-the-art alarm detection, evacuation and communications capabilities.

Developed in close cooperation with the Schaumburg Fire Department, the system monitors manual pull stations at every stairwell exit and elevator; duct-mounted smoke detectors on all fan systems; and ceiling-mounted smoke detectors and automatic sprinkler systems on all floors.

In the event of an alarm, the system automatically interfaces with the telecommunications system to call the fire department. A prerecorded emergency evacuation message is broadcast to affected building occupants by floor.

The system also automatically communicates with the BAS to adjust fan systems and dampers so that they provide smoke control by pressurizing floors above and below the floor with the fire. Stairwell doors which are normally locked from the stairwell side for security reasons are unlocked using this BAS interface. At the same time, elevators are recalled to the ground floor for use by firefighters.

A command center enables firefighters to achieve override control of elevators, fan systems, voice communications and smoke control devices. Equipped with a display panel which shows the location of the fire by zone, this center is also used to communicate with firefighters' phones which can be connected to jacks on each floor.

Another use of the BAS interface is to automatically produce color graphic displays related to the fire. Based on time of day, floor layout and occupancy level, including the number of handicapped people, these displays provide customized instructions indicating the most effective response to the fire. In addition, the BAS provides complete reporting capabilities for the fire management system.

TELECOMMUNICATIONS AND OFFICE AUTOMATION

Providing operating economies and technological access of tremendous value to tenants, telecommunications are based on the use of an integrated, voice/data, digital PBX system which is bit-transparent and software-driven. Capable of serving up to 4,000 lines, the system provides immediate access to a basic, feature-enriched PBX service via the use of multiline, electronic and standard analog desk sets, including executive workstations with CRT screens.

The PBX service provides all available phone features, including capabilities such as abbreviated or speed-dialing, last number redialing, call transfer, call forwarding, and three-way or multi-conference calling. In addition, the service offers a wide variety of high-quality, discounted long distance capabilities, including automatic least-cost routing.

The same digital PBX system used for telecommunications also provides tenants with a wide range of economical, technologically-advanced, office automation capabilities. The system can be used, for example, to easily and efficiently implement relatively high-speed local area networks both within offices and between offices on different floors. Providing protocol conversion, the system makes it possible to network virtually any office automation equipment while insuring that all devices will be able to talk to each other.

The system can also be used to communicate with off-site computers. In addition, based on tenant needs, the system can be connected to an on-site computer to provide tenants with in-house data processing, word processing and other data storage and retrieval services on a shared basis.

Rooftop satellite systems will also be offered on a shared basis, depending on tenant needs. A master antenna system, for example, will be used for video reception. Terrestrial microwave systems, paging systems and other types of RF-related transmission capabilities will also be made available as required.

Another significant economic and technological benefit of telecommunications and office automation operations is the ease and speed with which moves, adds and changes can be made. Both systems use the same universal and common wiring scheme based on five-foot modules and conventional, twisted-pair wiring. Every desk can therefore be wired to handle a phone, terminal or other device. And moves, adds and changes can be instantly made simply by plugging equipment into normal phone jacks and effecting software changes in the PBX system.

Business Support Center

Plaza Towers operates an on-site business support center which sells or leases telephone and office automation equipment and software to tenants. The center also provides tenants with on-site equipment maintenance, personnel training, temporary help and consulting services. In addition, a variety of other services are offered, including facsimile transmission, telex store-and-forward capabilities, and a message center operation which can be either personally staffed or automated in the form of an electronic voice mail system with broadcast capabilities.

CONCLUSION

Plaza Towers is designed to provide tenants with leading-edge capabilities in all areas of comfort, lighting, security, fire management, telecommunications and office automation. These capabilities are continually monitored, formally re-evaluated every quarter, and expanded and retrofitted as necessary to keep pace with advancing technology and the growing needs of tenants.

This helps to insure that tenants are able to operate with maximum efficiency and economy, not only now but throughout the life of the building. At the same time, it also insures that ownership reaps all of the benefits resulting from development of Plaza Towers as an intelligent building.

One of these benefits is that Plaza Towers has a competitive marketing advantage over other buildings because of its intelligent capabilities. This is enabling the building to be more successful in terms of leasing and releasing space. The building consequently is better able to achieve higher occupancy levels with the same tenants, resulting in a lower lease turnover rate.

Another benefit is that Plaza Towers makes it possible to increase cash flow through sales to tenants of telecommunications, office automation and other services. Cash flow is additionally enhanced by the capability of the intelligent building to charge-back tenants for off-hours use of facilities.

Plaza Towers is able to achieve lower occupancy costs because of the greater operational economy and efficiency provided by its building automation system. Heating and cooling levels, for example, can be easily adjusted to reflect changing environmental conditions at any time of day or night. In addition, diagnostic capabilities and operating data provided by the BAS insure that HVAC equipment is properly operated and maintained.

Finally, occupancy costs are further reduced through partnership in a venture with a systems integrator who has full, on-site, single-source responsibility for all intelligent capabilities and shared tenant services provided by Plaza Towers. Experienced and knowledgeable in all aspects of building intelligence, this company is better able to economically and efficiently achieve smooth, troublefree operations.

Improved occupancy levels, increased cash flow and reduced occupancy costs will all work together to create higher value for Plaza Towers upon sale or refinance. This enhanced value creation is the bottom line resulting from the development of the complex as an intelligent building.

Chapter 25

The Intelligent Health Care Facility

Kenneth M. Nottestad
Deaconess Hospital

The telephone rings in the office of the director of plant and properties at Deaconess Hospital in St. Louis. (See Figure 25-1)

"It's uncomfortable here on Four South," says a voice on the other end of the line. "Can you check it out?"

"Hold on," advises the director as he enters an inquiry on the keyboard of the telephone which is actually an integrated voice/data executive work station.

In seconds, a response to the inquiry flashes on the screen of the executive work station.

"It shows here that the discharge air on your air handling system is a little too high," the director tells the caller. "We will adjust it right away."

This is a voice/data communications procedure that has now become common at Deaconess Hospital. It is a capability that is the latest step in development of the medical institution as an "intelligent" health care facility.

HOSPITAL INSTALLED BUILDING AUTOMATION SYSTEM

A seven-floor, 550,000-sq-ft, non-profit medical institution with 527 beds, Deaconess Hospital completed a major building expansion program in 1979. (See Figure 25-2) With the expansion, utility costs increased by $225,000 in the first year. In addition, further large increases in electricity and natural gas

Fig. 25-1 Ken Nottestad, director of plant and properties at Deaconess Hospital in St. Louis, uses an executive work station as an input/output device for his building automation system.

rates were anticipated in coming years. Development of an energy reduction program therefore became of major importance.

In 1982, Deaconess Hospital investigated federal matching funds for energy conservation measures. The hospital applied for and received an energy grant award of $408,000 for four projects, one of which involved replacing an old, hardwired building control system with a new, microprocessor-based building automation system.

Six companies were interviewed and four were invited to submit proposals. Following evaluation of these proposals, final selection of a building automation system was made.

Price was important in the selection process but there were several other factors that outweighed price alone. One was that the system used an English-language format, eliminating the need for a numerical cross-reference book.

Another was the system's ease of use. This is important because the system is operated by shift engineers in the hospital's power plant. Consequently, to be properly used and accepted, it had to be user-friendly.

The local availability of experienced service and support for the system was also considered of prime importance. Furthermore, the system provided a remote CRT capability which makes it possible to monitor and control operations through a dial-up telephone capability.

System Provides Significant Benefits

Installed in 1983, the building automation system has provided significant benefits in monitoring and controlling the operation of two 500-ton chillers, three boilers and 23 air handling systems at Deaconess Hospital.

During its first year of operation, the system helped to reduce the hospital's annual electrical consumption from 15.7 to 14.9 million or by 800,000 kilowatt-hours. This is resulting

Fig. 25-2 Deaconess Hospital - North Building Facade

in a cost savings of more than $100,000 a year in total energy costs.

Operation of the system is also helping to increase personnel productivity by freeing the hospital's maintenance staff from routine tasks such as physically monitoring temperatures and controls. For example, the system automatically monitors chilled water return and controls chiller output temperature, eliminating the need for shiftmen to manually check temperatures and operate controls.

In addition, the system provides operating data which can be used to make more informed decisions in the area of HVAC (heating, ventilating, air conditioning) operations. Based on this data, for instance, it was possible to successfully implement a chiller optimization program.

System Performs Many Functions

Currently operating with more than 300 points, the building automation system performs a number of different energy management functions, based on individual loadsets established for Spring/Fall, Summer and Winter. These different loadsets are necessary because system operations vary with outside temperatures.

One energy management function is a demand limiting application in which the system monitors the kilowatt load in the hospital every 15 minutes and predicts what the load will be in another 12 minutes. Comparing this predicted load with a maximum, target load, the system begins turning off air handling units on a priority basis as necessary to insure that kilowatt consumption does not peak above target. The system also monitors space temperatures in areas where air handling units are turned off. If maximum or minimum temperatures are exceeded in these areas, the system automatically turns their air handling units back on while turning units off in other areas.

Another energy management function performed by the system is duty cycling. Air handling systems are automatically turned on and off each hour based on a preset schedule. However, many areas of the hospital such as surgery can only be duty-cycled during off-hours. Air handling units in these areas therefore are "locked on" during operating hours and "unlocked" only during off-hours. Space temperatures are also monitored as part of this function and overrides of duty-cycling operations are effected if maximum or minimum temperature values are exceeded.

A programmed start/stop energy management function permits air handling systems in certain, non-patient areas to be totally turned off during unoccupied times. But this function is also subjected to space temperature overrides as required.

A chiller optimization program operated by the system monitors chilled water return temperatures from three different points. Based on an average of these return temperatures, the system automatically controls chiller output temperature, reducing the amount of chilled water produced to a minimum.

Temperature alarms also are maintained and monitored by the system throughout the hospital. In addition, the system is used to monitor run times of all major HVAC equipment. This makes it possible to schedule preventive maintenance based on actual hours that equipment is run, rather than simply on the basis of calendar scheduling.

DIGITAL SWITCHING SYSTEM IS INSTALLED

Following implementation of its building automation system, Deaconess Hospital installed a new digital telephone switching system with integrated, voice/data communications capabilities. This system upgraded the hospital's telecommunications capabilities by providing a host of special features such as call-waiting, call-forwarding, permanent holds, consultation holds and three-way conference calling. In addition, the system can be used to achieve other capabilities such as least-cost routing in the future.

Deaconess Hospital then spearheaded development of an interface between the digital telephone switching system and its building automation system. This made it possible to communicate with the building automation system as well as the telephone switching system using interactive video display terminals, the integrated voice/data workstations.

Hospital Uses Three Workstations

Currently, Deaconess Hospital operates three integrated voice/data workstations. As mentioned, one of these workstations is used by the director of plant and property to monitor and control operations of the building automation system.

Using the workstation, the director can answer all questions received from building occupants without having to bother personnel who operate the building automation system at the power plant. The director can also input data to the system, control existing system operations or program new operations.

In effect, any function that can be performed on the keyboard of the building automation system can also be done at the workstation. Furthermore, the workstation can be operated at the same time that someone is working on the system keyboard without interference. In addition, the workstation can be used equally well in the home as well as the office of the plant and property director.

A second workstation is used to enter telephone moves, adds and changes into the telephone switching system. A third workstation is installed in the home of the manager responsible for the telephone switching operations, enabling him to remotely diagnose and solve problems at night or other times when he is not physically present in the hospital.

FUTURE PLANS INCLUDE FIRE, SECURITY

Currently, Deaconess Hospital provides programmed start/stop capabilities to an adjacent medical office building as well as its own facilities. In the future, the hospital is considering use of the system to perform duty cycling and energy management functions for a nearby nursing home called Deaconess Manor.

Also under consideration are plans to upgrade fire management and security monitoring capabilities. Presently, for example, fire management is handled by means of a hardwired system with about 150 alarms which are coded to indicate approximate alarm locations.

This system could be upgraded by increasing the number of fire alarms to about 200 and conducting fire management operations using the computer-based system. Interfaced with voice/data workstations, this system will make it possible to automatically and instantly identify the exact locations of alarms.

Security monitoring will involve the installation of intrusion door alarms in key areas such as the central store room, pharmacy and loading docks along with the medical office building and Deaconess Manor. The computer-based system will be used to program these door alarms so that they would be activated at preselected times such as midnight to 5 a.m. At the same time that alarms are activated, the system would printout the time and location of each alarm incident.

Overall, the administration and board of directors of Deaconess Hospital are conscious of the need to operate ever more economically and efficiently. Consequently, they are supportive of efforts to develop the hospital as a building that meets criteria for possessing "intelligence."

Chapter 26

Computerized System Key To Intelligent Preventive Maintenance Management

Peter Robinson
Director of Operations
British Columbia Housing Management Commission

The British Columbia Housing Management Commission is significantly reducing its HVAC operating costs while increasing tenant comfort by using a Computer-Aided Maintenance Management (CAMM) program to "intelligently" coordinate its maintenance contract.

The CAMM software program produces preventive maintenance work orders and repair orders. But more importantly, from those orders it produces a series of management reports which the Commission uses to supervise a multi-year maintenance contract with an outside contractor.

Based on the first years results, the Commission is optimistic about reaching the goal of the program -- to reduce annual repair costs from $150,000 per year to only $40,000 per year in three years. It has already experienced an average decrease in HVAC work orders of 25 percent in the first year. Both the total volume of work orders as well as those primarily concerned with HVAC have declined.

An outside contractor maintains the physical plant equipment for more than 90 percent of the 213 buildings managed by the Commission in the Greater Vancouver metropolitan area. These buildings, located on 31 diversely located sites, consist of close to 6,000 provincial and federal-provincial housing units.   The public housing units are for low-income families, the handicapped and senior citizens.

Physical Plant Initially In Poor Condition

The physical plant equipment in the buildings consists of primarily boilers and air handlers. More than 1,500 pieces of equipment are logged in the CAMM program. Individual record keeping is automatically referenced by each individual piece of equipment as well as by the building.

In late 1983, the Commission began a study to examine the growing volume of maintenance work orders. It looked at all the incident reports, the damage reports, downtime reports, and the major tenant complaints. This study showed that the Commission was not handling the key systems of heating, ventilating and controls very well.

Repair costs were quite high. A fair amount of money was being spent piecemeal each year on basically unplanned plant maintenance. Plus, at that time, no single outside firm was maintaining more than one building. No one had a firm idea of the total scope of what was being done.

The study reviewed these costs of outside contractor work and the costs of its own maintenance people who were being taken away from their primary responsibilities such as repairing broken doors, faucets, toilets, and mirrors.

The results of this study suggested that a single, comprehensive contract with an outside contractor would be less expensive than what the Commission was currently paying for numerous contractors and in-house maintenance staff time on physical plant related problems. The Commission foresaw the possibility of reduced costs and a better product through such a contract.

Single, Large Contract Necessitated CAMM

To facilitate the bidding process for a single comprehensive maintenance contract, the Commission inventoried the entire physical plant in all the buildings. A consultant was then hired to help draft the contract specifications. This included the sometimes difficult task of determining the fine-line between in-house staff responsibility and the contractor's responsibility.

The scope of the contract essentially includes all the equipment in the physical plants and mechanical rooms. The Commission maintenance staff does the work on the piping and ducting throughout each building. They are also still the first to respond to a problem. They determine whether the problem is in the physical plant or some other part of the building.

But the Commission did insist on the CAMM software program to manage the maintenance contract. Such a system impressed Commission management because it would provide them with the ability to continually monitor the current status of each building's maintenance. They could see the benefits of CAMM's historic record-keeping in preparing a three-year capital maintenance program. It provides an estimated lifespan of equipment, energy conservation reports and a number of other management tools.

The Commission also insisted on a CAMM software program for managing the maintenance program because it increased the confidence factor in selecting a single firm for a contract of this magnitude. It helped ensure that the contractor chosen would be a professional firm with extensive experience handling large contracts. Typically, firms of this caliber have already developed these types of programs to control their own operations and for sale to their customers.

After these contract specifications were developed, which involved extensive negotiations between all the parties, there was a public bid offering and the lowest bidder was selected.

Goal To Cut Repair Costs Two-thirds

The Commission knew that the plants were in poor shape. Thus it built into the contract repair program to get the buildings back into normal operating condition so an outside contractor in conjunction with the CAMM preventive maintenance program could maintain them. This repair program consists of identifying a deficiency list of repairs required to get the existing physical plant to the point where the Commission can save money.

The process of getting the buildings up to a maintainable level cost $150,000 the first year. But using the CAMM program, the Commission has an agreed goal with the outside contractor to cut those costs IN HALF the second year to about $75,000, and then IN HALF AGAIN the third year to about $35,000 the third year. By the fourth year, the goal is for repair costs to drop down to whatever preventive maintenance can maintain it at--which is estimated to range between $25,000 to $40,000--indefinitely.

Based on the results thus far, the Commission is optimistic that those goals will be reached. The other benefits are reduced energy consumption, reduced tenant complaints, and reduced work orders. And with the reports from the computer system, the Commission can watch the savings accrue over the years, which will enable it to prove the success of this three-year contract.

Of a personal concern to the Commission were the previous emergency maintenance costs. No heat is an emergency, especially for tenants like senior citizens who rely on that comfort level and constant delivery of heat. The buildings were reaching an age with the original heating systems where the life of certain equipment was expiring. But as a result of these improvements in the way it does maintenance, the Commission reported a reduced number of tenant complaints, better quality of service and comfort, monetary savings and a longer lifespan of the plant equipment.

In-house Staff Acceptance

The in-house maintenance staff for the Commission not only has more time for its regular duties now, but staff members also have time to learn new skills like dry-walling. Technicians for the outside contractor also hold in-house seminars with the Commission maintenance staff members to discuss the continued importance of their working knowledge of the system in conjunction with its services.

The Commission maintenance has three maintenance managers for each of the regions in Greater Vancouver. The regions are Greater Vancouver Burrard, Greater Vancouver Central and Greater Vancouver Fraser. There are about fourteen maintenance people for the three regions combined.

Before the present contract, the maintenance staff reported the effects of having little continuity between the various contractors performing maintenance and repairs. One contractor would sometimes reverse what the previous contractor had done. Each one liked to blame the previous contractor.

The CAMM program reports are of great assistance to the regional managers. Before it, they never fully knew the current maintenance status of their buildings. Without the computerized reporting system, they had to rely more on informal communications.

CAMM now produces a tabulation of the frequency of calls, the type of calls, the sites of work orders, the dates of actual maintenance and a brief description of the activity performed. (See Figure 26-1) This report is automatically produced for the regional managers on a monthly basis with the custom report generator. These reports are not labor intensive, and they provide monthly feedback to the regional managers about the types of problems in their buildings. CAMM requires about two hours a week to input each week's information.

Instead of expanding the in-house staff with a variety of specialized mechanics whose expertise probably could not be utilized 100% of the time, the outside contractor provides the Commission with the services of qualified journeymen, electricians, control fitters, and boiler mechanics with the appropriate "gas ticket" permits. The CAMM program schedules specific trades to perform just that type of work only when they are needed.

Fig. 26-1   The TRIM computerized maintenance management system produces the necessary work orders and customized reports for the British Columbia Housing Management Commission. Larry Bell, manager of technical services, and Peter Robinson, director of housing operations, both of the Commission, discuss one of these work orders with Audrey Rinke, Johnson Controls service coordinator and dispatcher.

Advantages of Improved Technology

As a result of new controls technology, public housing today provides a better standard of living than it did previously. The Commission is providing control systems that work more quickly and reliably.

A case in point occurred in the middle of June one year when Vancouver experienced unusually cold, wet, damp weather almost like winter which was very hard on the senior citizens. In the past, heating systems were shut down during the summer, and it took one or two days to restart them. Now with modern control systems, heating systems are ready during the summer. Sensors and controls are more sensitive and eliminate problems like those caused by sudden winter-like weather in the middle of June.

Administrative cost are also decreasing, which is usually an indirect cost that managers do not see. Because the Commission does not have to process the large volume of records, logs and summaries--the computerized maintenance management system does that -- clerical time can go toward other projects. Before, the regional managers were requesting extra clerical staff just to process reports.

One single maintenance contract is also easier, and perhaps only possible, with the CAMM program. Senior management receives copies of the reports too so they can now better manage regional manager's skills and abilities. They can tell regional managers what is going on in their buildings rather than vice versa. It is a terrific appraisal tool for senior management to assess their staff productivity.

Energy Conservation

The Commission is already implementing some energy conservation measures at certain test sites that it unveiled during its new maintenance program. It estimates that retrofit of nine buildings has resulted in an average annual savings of $2,300 per building. These energy conservation measures consist of standard practices, such as installing controllers for boiler sequencing, radiation scheduling, night setback of temperatures, programmed start/stop of equipment and heating plant shutdown based on outdoor air temperature.

Return on investment is typically less than one year for these retrofits. These savings may seem minor, but when multiplied by the number of buildings that the Commission manages, they can quickly become significant. Its plan for the next three years is to retrofit 100 more buildings with an estimated savings of $2,500 per building for about $250,000 total savings.

Future Benefits Of CAMM

With a proven record of fiscal responsibility provided through the documentation of the CAMM program, this will help justify budget requests for capital maintenance in the future. This program indicates that the Commission is getting control of these costs.

Another of the advantages of the CAMM program in the future is that it will give the Commission suggestions for capital maintenance improvements and retrofits both from an energy point of view and an operating point of view. This is especially true in the controls area where technology is taking off.

The Commission takes pride in its utilization of the latest technology in monitoring and performing preventive maintenance for the Greater Vancouver housing properties. It wants to make this the most cost-effective type of housing it can. This is part of its plan to be fiscally responsible to the citizens of British Columbia as well as to the housing tenants. What is most important is that it is achieving these savings and at the same time increasing the level of satisfaction for the tenants.

Chapter 27

Building Intelligence Helps Houston Office Complex Stay 95% Occupied

Giorgio Borlenghi
President of Interfin Corporation

At a time when Houston is making national headlines for its empty office buildings, Four Oaks Place has an occupancy rate of more than 95% -- one of the highest in the city and one that would be considered good anywhere.

As the developer and manager of Four Oaks Place, Interfin Corporation attributes its success to a number of factors including the decision to install a computerized building automation system to monitor and control mechanical equipment. In the often misunderstood world of "building intelligence," this type of system is a key building attribute.

The building automation system helps to lease the four-building complex by providing comfortable temperature and humidity conditions for tenants and by reducing costs for electricity and labor. It also helps to impress tenants and investors for Four Oaks Place and other projects with Interfin's operations. The system capabilities are currently being marketed to other buildings in Houston.

Four Oaks Place is located in the Post Oak area of western Houston. The four buildings -- a 30-story central tower flanked by two 25-story towers and a 12-story mid-rise building -- provide a total of 1.75 million square feet of office space.

Computerized System Attracts Tenants

The primary benefit from the building automation system is that it helps to attract major tenants to the project. In the initial leasing, the computerized system helped to solidify deals. It was a contributing factor with some of our major tenants.

Most large tenants will be in a project for 15 years and are, in essence, paying the electricity bills as well as other bills. They like to look at how Interfin operates the buildings and at the budgets; and they compare them to the market through real estate brokers, consultants, or industry standards.

One of the major tenants at Four Oaks Place brought in an outside architectural/engineering firm that reviewed this project. They looked at the operations, at the computerized energy management; and they gave their blessings to how the plant is operated.

System Reduces Developer's Costs

Interfin Corporation's cost for utilities at Four Oaks Place is now over $2,000,000 per year in 1987. It would be more if the building automation system had not been installed.

In the old days, all a developer/owner did was build a building, put in some time clocks and let it run as the tenants paid all the electricity. But now, with the free rent concessions, the developer/owner must pay for some of the electricity. So he's taking a more professional and concerned approach to what type of energy management systems go into buildings -- more so than he did a number of years ago when electricity was inexpensive and buildings were more fully occupied. Utility costs of more than $2,000,000 per year lends credence to buying a computer for energy management.

System Impresses Investors

A major project should utilize the highest quality products to be rated Class A. Interfin made the The Four Oaks Place operation as first class as it could.

Investors and potential lenders have toured the plant and have seen that they would be investing not only in the buildings and leases but also in exceptional mechanical equipment. They have been amply impressed with these operations.

Interfin explains to investors how the building automation system operates the plant in an optimum fashion and reduces expenses from a utility standpoint and from a manpower standpoint. The labor costs are lower because any mechanical problems are identified, and often corrected, at a central location where the system controls all four buildings.

System Helps Developer Get Other Business

Interfin also benefits from using the building automation system to show the quality of its operations to potential clients for other projects and from marketing its capabilities to existing buildings in the area.

From a corporate standpoint, the company also uses it as a resume piece for new tenants who want Interfin to build a building for them. The system is one of the reasons for the success of the company, so Interfin shows them how the building operations staff programmed the computer and designed the system to benefit the project.

The building automation system is helping the company enter the fee management business. The computer is capable of tying more buildings into the system over telephone lines, so Interfin can manage a number of other projects. It has made the front end investment and has a lot of programming ability as well as operating experience. With the manufacturer's help, it is ready to actively market the service of

Fig. 27-1   John (Williams) Berry III, director of engineering for the Interfin Corporation, operates a building automation system that contributes to an office occupancy rate greater than 95% at Four Oaks Place in Houston.

monitoring and controlling mechanical plants in other buildings in the area.

Central Plant Monitored and Controlled

The 5,500-ton central plant at Four Oaks Place provides chilled water for the entire complex. The chillers and auxiliaries of the cooling system, as well as some 52 various pumps, electric heating and the variable air volume system, are centrally monitored and controlled by the building automation system.

Interfin also is able to use the system from off-site through telephone modems so that it is not necessary to staff the plant 24-hours-a-day. In the event of a problem after hours, the system sends an alarm to the security station. Security staff then contacts an engineering staff member who is able to call the building automation system from his terminal at home to diagnose and often correct the problem.

System Saves 20-30%

Various energy management programs have been utilized over time. Installation of the system during construction so that air conditioning could be provided as needed and equipment could easily be turned on and off saved the developer money. Installing the system resulted in an estimated 20 to 30% overall savings on electrical consumption. Programs utilized include chiller manager, chiller sequencing, condenser water reset, chilled water reset and optimal run time. Demand limiting is done in winter for the electric heating system, which has held down demand charges without sacrificing tenant comfort.

Commitment Is To Tenant Comfort; DSC's Help

While Interfin operates the complex as efficiently as possible, it is also responsive to tenants. It must conduct business

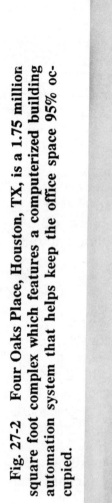

Fig. 27-2   Four Oaks Place, Houston, TX, is a 1.75 million square foot complex which features a computerized building automation system that helps keep the office space 95% occupied.

in a good atmosphere and HVAC is part of that environment. Tenant comfort is something Interfin is committed to.

Interfin uses digital system controllers as field panels for the building automation system because of the developer's concern for tenant comfort. When it bought the digital system controllers, the specifications required back-up programs in the panels for all the air handlers. Although the central computer has been real dependable, if it were to go down, the system would lose some energy programs temporarily but the tenants' comfort would not be sacrificed. The air handling units would continue to turn on and off as needed without Interfin having to do anything.

The field panels are smart and can be programmed for many things. Interfin plans to use them for more distributed processing in the future.

A computerized building automation system is a valuable tool for Four Oaks Place. It is hard to put a price tag on all the benefits from one.

SECTION VII

Developing Tomorrow's Building Today

Chapter 28

Measuring A Building's IQ

James Carlini
Carlini & Associates

There has been a great deal of interest in developing the concept of sharing communication and information technologies in a building among real estate developers as well as high tech vendors. Some developers have taken the plunge and have introduced new buildings that have demonstrated varying degrees of intelligence as well as success. Others are waiting to see if there is a cookbook approach to developing and retrofitting buildings into smart buildings that will attract tenants in both stable and overbuilt markets. Up to this point, there has not been a way to clearly define what intelligence is, let alone compare intelligence among competing buildings in the same geographic market (See Figure 28-1).

Today, there are no recipes for success. Combining communication services, word processing centers, and other information technologies into packages that tenants can opt for is still in its early development. Many shared tenant service providers are still in an experimental stage of offering a mix of services without fully understanding the unique needs of each building. The difference between an expert and a novice in this fledgling industry is one building. As more projects are undertaken, the providers will identify common trends across geographic markets and offer packages to the developer that reflect experience and not hypothetical design.

This concept of creating a smart building is not limited to new construction. Several projects have already been undertaken to upgrade an existing building by rewiring it and install-

Fig. 28-1   How Do You Measure Building Intelligence?

ing the communication systems and information technologies for shared use by the tenants. This type of project is called a retrofit and is considered to be more difficult to implement in some aspects than new construction. In either type of project, the ability to measure what is being installed in the building let alone the tenant services being established is very desirable, yet not common.

In a recent survey, the author discovered that not all users knew what shared tenant services actually meant. Many requested a definition of what shared tenant services would provide and how it made a building intelligent. "How do we know what one intelligent building offers compared to another" was a typical response.

Another finding of the survey was that shared tenant telecommunication services was of interest to some large companies that had remote offices around the country. Corporate telecommunication managers felt that it could be a way to alleviate some of the problems associated with evaluating, selecting and maintaining a phone system for a remote office by having them simply subscribe to the building's system. Prior to this response, it was felt that shared tenant services appealed only to the small and medium-sized businesses found in a multi-tenant office building.

The proponents of the intelligent building concept, which at the time of this publication is only three years old, have to educate a huge target market. At this point in time, some people would say they are not doing too good a job. Many individuals in the real estate business are not comfortable with all of the intelligent building terminology, let alone the marketing approaches. Many vendors have yet to come up with an approach that satisfies everyone's concerns. Many developers are hesitant to add shared tenant services to their new projects. They are waiting for the market to shake out and for the quality providers to emerge.

One thing is very apparent: there are many big companies viewing the intelligent building industry as a new service in-

Exhibit 28-1   BUILDING INTELLIGENCE

Level 0   Provides non-computerized control for energy management and elevators. It may or may nmot have any life/sfety or security systems present. (This building has no intelligent amenities and does not qualify for an intelligent building classification.)

Level I   Provides an infrastructural core, including computerized systems for energy management for Heating, Ventilation, and Air Conditioning (HVAC), elevators, life/safety and security systems. (This building is The Frankenstein.)

Level II   Provides tenants with Level I capabilities plus share conference/meeting room space, photocopying facilities, word processing centers.

Level III   Provides tenants with Level II capabilities plus telecommunications services including long distance voice, some data communications and basic plain old telephone service (known in the communications industry as POTS) utilizing its core cabling system.

Level IV   Provides tenants with Level III capabilities plus sophisticated office automation and information processing services including wide-band communications for video conferencing, high speed voice/data communications. (This building is the Einstein.)

dustry that is here to stay. There has already been some shakeout in the industry, but it has not deterred new entrants into the market. The pricing is evolving from a discount type of service to a premium service due to convenience and high quality service provided to the subscribers. Marketing intelligent building concepts creates the need for a multi-level approach to convince the following individuals that an intelligent building offers more than a conventional building:

Developer (Owner)

Financier (Appraiser)

Leasing Agent

Facilities Manager

Tenant

The above-mentioned individuals must have a good understanding of intelligent buildings in order to evaluate the shared tenant services (intelligent amenities) offered in various buildings. The benefits, as well as the risks, of providing or subscribing to shared tenant services should be clearly understood.

The term "Building Intelligence" has to have some definition to it. Office building intelligence can be divided into one of five categories depending on the level of intelligent amenities present in a building (See Exhibit 28-1).

Level I capability is viewed as a basic requirement by most tenants, facilities managers, leasing agents and developers. The higher levels of building intelligence are not currently understood nor appreciated by many tenants, leasing agents, appraisers, financiers, and developers. One could conclude that as the level of building intelligence increases, the awareness and acceptance decreases from all of their perspectives.

If we analyze each individual's perspective, we see that the developers, or owners, should be able to measure the surrounding competitive buildings to determine what services should be provided in their building to attract and maintain tenants. In some cases, developers have implemented too many services or the wrong mix of services and have not generated the anticipated revenues that were initially expected (See Exhibit 28-1). As time goes on, the developer will be able to be more successful in launching intelligent buildings due to the fact that measuring what the competition offers will become more of a science than an art.

Measuring building intelligence from a financier's perspective is critical. The financial organizations have the greatest amount invested in the actual building, yet they are not focusing on the critical issues of applying communication and information technologies that affect value in newly constructed and retrofitted intelligent buildings. If appraisers use the standard income approach in establishing value in a property appraisal, they should include the revenues generated by the intelligent building services and any roof-top communications revenues that the property generates. By including this income stream in the appraisal, the intelligent amenities along with the roof-top revenues will change the whole evaluation of a property. As more and more financial institutions become comfortable with this expanded approach, the developer that is shopping for capital will have to compete with those developers that include these new revenue streams on development projects that potentially reduce the risk of the lender and creates a "safer" project to invest in.

The leasing agent's perspectives on measuring intelligent buildings are ones that have to expand their understanding of the terminology as well as the comparison of different intelligent buildings for prospective tenants. If leasing agents do not understand the additional benefits afforded to a tenant in an intelligent building, they will not be able to provide good service in helping the tenant select the best office space suited for the tenant's individualized needs. The leasing agent must become comfortable in analyzing a whole new set of amenities as well as being able to differentiate the services provided by one intelligent building from another. In today's market, the leasing agent is not exploring the advantages and selling points of intelligent amenities to their fullest extent.

In any office building market, the tenants' perspectives focus on critical elements that attract them to certain buildings. All tenants looking for space focus on certain minimal amenities including:

- Location
- Heating, Ventilation and Air Conditioning (HVAC)
- Security
- Parking
- Building Maintenance (Janitorial) Services
- Elevator Service (in major office buildings)

They are concerned with the level of service as well as the level of comfort in an office setting. With greater dependence on communications and information technologies, tenants also need support in putting together the right mix of these systems and services to allow their businesses to maintain a competitive edge.

As one can see, the traditional elements that attract high quality tenants are beginning to evolve and expand. With that expansion into these new communication and information technologies, the overall quality of a building and its property must be evaluated with an analysis that includes these new amenities.

Up to this point, most tenants and leasing agents did not have the level of awareness to take these amenities into consideration when evaluating office space. A tool providing the prospective tenant, as well as the leasing agent, with a quantifiable way to compare different buildings and focus on this significant factor was needed.

There are other factors that are considered in the decision-making process that already have a high-level of awareness including location, comfort, and others that vary greatly with the organization and its individual preference. The ability to objectively compare buildings and their intelligent amenities through a comprehensive test that measures the amount of services and systems present in a building has become significant in this decision-making process.

This chapter's author, James Carlini, created the Carlini Building Intelligence (CBI) test in 1985 that focuses on measuring a building's IAQ (Intelligent Amenities Quotient). (See Figure 28-2). He designed the information technologies portion of the test and J.M. "Jack" Fischer, Manager of Automated Systems for Urban Engineering developed the Mechanical, Electrical, & Automated systems portion. Together, they designed the test to consider all critical operations in an intelligent building.

The CBI test is divided into three parts: communications, information technologies, and building automation technologies. The communications portion consists of:

- Telecommunications (Voice, Data and Video)
- Local Area Networks (LAN)
- Cabling Distribution
- Roof-top Communications
- Security/Elevators

The information technologies portion includes:

- General Information/Building Characteristics
- Office Automation
- Information Processing
- Document/Records Management

The building automation technologies portion includes:

- Environmental Systems
- Temperature Control Systems
- Building Automation Systems
- Life/Safety Systems
- Lighting Control Systems
- Elevator Systems
- Emergency Power Systems
- Building Operation Radio Systems

The intention of the test is to bring a more informed audience to the negotiations table than what has been the case in the past. It enhances the tenants' information base and provides a yardstick to measure areas that have been previously vague or non-existent. The test can also be viewed as a tool to the appraiser that is trying to evaluate a building and take into consideration these new intelligent amenities during the evaluation process. To the developer, the test provides a means of measuring the market to determine what services and amenities should be established to remain competitive with the surrounding developments.

The CBI test has been used to measure and compare different Class A office buildings that offer similar intelligent amenities.

It provides a way to separate the "Einsteins" from the "Frankensteins" by giving a numerical score for each building so that there is an objective way to differentiate buildings, even when all of them claim to be the most intelligent. (See Figure 28-2)

The test provides the prospective tenant, leasing agent, developer or appraiser a definite score on what the building offers in the way of communications, information, and building automation systems. The use of those scores can help

Fig. 28-2   In markets where most developers offer intelligent buildings, there must be a way to quantify the buildings' IQ*

*Intelligent Amenities Quotient

everyone understand the market, the pricing of leases, and the differences between intelligent and not-so-intelligent buildings without having to be an engineer or a computer scientist.

The tenant, as well as the appraiser, must also consider factors that include energy costs per square foot per year, service costs for long distance calls per minute and other service costs for comparisons of different buildings as well as outside versus inhouse services.

As the test results would support, a label of "intelligent building" does not mean the least expensive nor the most expensive building to operate. It designates that certain services and technologies are available for the tenant. The tenants must decide whether or not those services fit the needs of their businesses. The mere fact of having these intelligent amenities in a building does not discount the importance of the tenant understanding and taking into account the price performance of services in a smart building.

One thing is for certain, state-of-the-art tenant needs will not be satisfied by turn-of-the-century building amenities. As many organizations are realizing, we are currently shifting from the Industrial Age to the Information Age. The needs of office building tenants, who are really providers of products and services themselves, are being shaped by domestic as well as international pressures to be more competitive and to maintain high quality standards. These needs are evolving to focus on intelligent building amenities that were not important, nor existent, in buildings that were constructed as little as ten years ago. The shift in needs has created a new dimension in applying communication and information technologies at a building level rather than an individual tenant level to realize the economies of scale. This shift has established what most industry experts commonly refer to as the intelligent building concept.

When the intelligent building concept explodes into all of the different geographic office building markets, a way to objectively compare new and existing buildings and their amenities in both stable and overbuilt markets will become critical to owner, leasing agent and tenant as well as to the financial institutions that have major interests in these development projects. Being able to measure building IQ in order to ask and answer the question "How smart a building do you have?," is as easy as "How much space to you want to lease?" will give everyone working with real estate a much higher comfort level in both selling and leasing high tech office space.

Chapter 29

Integrating Intelligent Building Technologies:

The "Chief Information Officer"

By John A. Bernaden
Johnson Controls, Inc.

In compiling and editing the chapters for this sourcebook, one consistent message came through: the technology for intelligent buildings exists today.

Certainly, there remain open issues, such as a precise, industry-accepted definition of an "intelligent building." There is also no IQ formula for industry analysts to gauge just how "intelligent" a given building is. But, as you have read, this sourcebook clearly demonstrates that intelligent building technology is available and being applied.

This rapid technological development, however, has outpaced the evolution of management structures needed to take full advantage of the intelligent building systems. While the technologies have grown more sophisticated and more integrated with other building and business systems, management structures often remain provincial and fragmented. A similar, related situation is taking place within corporate information structures. There, a position of Chief Information Officer is slowly evolving to give the needed direction to information systems for corporations. The intelligent building industry would be well-advised to consider this development and its implications. In the preliminary, positioning process, managers of data processing and telecommunications departments are already lobbying for the role of future CIO in an effort to control *all* information-gathering, transmission and processing activities within an organization or campus.

Though spoken of as all-encompassing, the CIO concept pays little heed to building management concerns. It therefore threatens to leave facilities managers and those in the intelligent building industry with subservient roles, as key equipment and many processes of intelligent building systems become integrated under data processing and telecommunication control.

Facilities managers and those in the intelligent building industry should not sit idly by as their prerogatives are eroded by other departments. They too must assert their positions as controllers of strategic information systems that are integral to the performance and, ultimately, profits of business. Only then they will have influence over the evolution of the CIO and -- ultimately -- the future of intelligent buildings.

The Need For Information, Integration

America is entering the "Information Age," as award-winning author and public speaker John Naisbitt aptly suggested in his blockbuster best-seller, *Megatrends*. Productivity in this new age depends on a business's ability to produce and utilize information. That information is then a crucial element in forming business strategies, which have a long-term impact on business performance and profitability.

Information, however, must be differentiated from raw data or simple facts. It is the task of low-level technicians, using everything from a calculator to mainframe computers, to collect data on business and building conditions. That data may have tactical applications for field personnel. But the management levels require analyzed data -- pertinent and timely information -- to direct strategies for operating a business. Without that information, upper management is prepared only to draft strategies and make decisions in response to gut reactions, business crises or unrealistic expectations.

In that context, intelligent building systems are a part of the mechanism for supplying data and ultimately information. Simply put, an intelligent building is an information system designed to improve tenant's productivity. For example, the building's facilities management system is a significant cog in the machinery. The building's data processing capability, likewise, is another pivotal point. The building's telecommunications is then the conveyor belt for data and information distribution in the process. Technologically, these three groups have been converging, as seen in some of the most advanced intelligent buildings. Facilities management technology involves computerized collection and processing of data and information. The data and information are transferred, quite often over telephone lines, using modems or other interface equipment. Telephone and building signals even exist on the same fiber optic cables in some instances. Building information is being readily transferred into data processing computers that are used for such activities as billing tenants.

A synergistic relationship is forming. For example, data processing and facilities services departments have worked together in selecting and implementing computerized maintenance management systems. The data processing departments better understand the specifications necessary to select the proper computer hardware and programs. The maintenance staffs understand physical plants and what needs to be programmed.

In this way, in a growing number of instances, they have effectively helped each other procure and set up the system that will best benefit everyone in the organization. Once this was done, the maintenance staffs were able to more efficiently schedule and manage preventive maintenance of the physical plant equipment. In a select few large corporations, the central mainframe computer itself is serving the dual functions of data processing as well as energy management. Local area networks are also revolutionizing how computers of all sizes will soon share information.

Technological advances are prompting the need for further cooperation. Information processing continues to require more sophistication and integration, as networking capabilities mature. Reliability of the networks increases, while the costs continue to decrease. The focus should shift toward sharing the elemental data and resulting information among data processing, telecommunications, facilities management, and factory automation departments.

But such cooperation assumes an organization with interdisciplinary communications working towards a single set of well-defined goals. That makes good sense, and few would argue against it. Unfortunately, in reality there are few organizations with such far-reaching information strategies, much less clearly-defined ones.

Information Systems Require Strategic Plans, Management

A speech by Bank President Thomas S. Johnson at the Financial Executives Institute exemplifies the situation. He said that one would expect that Chemical Bank, with cumulative spending on technology of one billion dollars over three years, would have a specific picture showing where each expenditure would fit in and contribute to the organization. But it didn't. Johnson has helped change that, as Chemical Bank moves towards greater strategic planning over its information systems.

The strategic planning phase for information and automation systems is missed altogether in most businesses. Once a technology is adopted and in use within an organization, the system is generally overlooked in development of information strategies by top-level management. Johnson's own analysis pointed to the problem: "Most of the technology investments were generated from the bottom up, group by group.... We had major expenditures for systems that were directed by people at the most junior levels of the organization." In that situation, one group cannot know what the others are doing; redundancy cannot be avoided; Return On Investment, not corporate goals, becomes the measure of success.

Coordination, direction and integration are needed. Someone will ultimately have to take charge: someone like a CIO. The CIO would ideally coordinate and direct personnel and the systems that provide top management with the necessary information for strategic planning. As a member of upper management, the CIO would have the essential broad view -- seeing exactly what information top management must have and the technical means available for getting that information. But an equally important responsibility of the CIO should be the creation of an environment where those strategic plans could be implemented. The CIO would coordinate and direct all departments responsible for developing the most productive, intelligent work environments possible.

For example, this means providing a flexible, efficient workplace. In the office of the future, a single integrated terminal may sit on the average office worker's desk. Included in it would be a telephone, video display terminal, keyboard, environmental sensors such as a thermostat, humidistat and air quality/ smokestat, along with security sensors such as a biomedical scanner to identify the operator. All information would be transmitted in and out of this single, compact terminal via an ISDN cable. With convenient universal outlets located throughout the floor space, relocation of the terminal would simply involve moving the unit to a new location and reconnecting it there. Cubicles and office dividers in the future will be more flexible and integrated with lighting and cabling designs. Thus, as they are moved too, the lighting and cables can be easily rearranged to maintain an environment designed for maximum productivity.

Early models of integrated voice-data terminals (IVDT) are already available and are becoming very popular. Flexible workplace environments are similarly a sign of the times. Today, hundreds of corporate efficiency experts are studying how to improve the productivity of America's workforce and the quality of American products. In general, they have found that much of the problem is *not* with *American workers* -- it is the way American management makes workers perform their jobs.   An intelligent office environment and information processing systems will help improve the way people work. With a single corporate viewpoint, the problems of redundancy or mismatching systems can be avoided.   That would not only facilitate collection and analysis of information, it would also help control costs for installing and operating the systems.

But the CIO concept has been catching on slowly.   Organizational structures and dynamic politics involved in creating and administering an intelligent information environment, never mind an intelligent building, have so far impeded the concept's development.   The organizational segments found in intelligent buildings -- like facilities services, data processing and telecommunications -- have developed their own "turf" of technical expertise.   Formal management for each of these segments is a relatively recent phenomenon developed in the last half of this century.   As with siblings who covet their own belongings, the synergistic benefits of working together have eluded these separate groups.   Cooperation often seems accidental or grudgingly forthcoming.

A recent case in point found a telecommunications manager facing prohibitive costs for trenching under a busy street to run a cable to another building.   Desperate after exploring every technological avenue, he discovered an existing wiring conduit used for building controls.   Unfortunately, he was told by the facilities manager that it was already too full.   The telecommunications manager then explained how the planned fiber optic cable could easily handle the information requirements of the building controls and still perform its telecommunications functions. Running the fiber optic cable through the conduit saved the telecommunications department from the prohibitive costs of the trenching.   It also expanded the capabilities for additional building controls transmission.

This seemingly obvious point is that communications and cooperation had not taken place at the beginning of the project.

A cohesive, well-planned approach to the information systems would have avoided the debate and possibly, through foresight, would have avoided the problem altogether. A CIO would have been able to make a decision immediately, in accordance with the long-range strategies for the company's information systems.

Telecommunications, Data Processing Take Charge

Many business now look to telecommunications and data processing for expertise when drafting and implementing information system strategies. They are leading the way towards creation of the CIO because they have adroitly placed themselves in positions of control over the many technical and managerial choices required for information system strategies.

The move toward the CIO concept is further evidence of an on-going, natural selection that takes place as technology and business environments change. Business organizations themselves evolved from having office managers in charge of nearly everything. The spread of new technology and systems then created new specialties in business management: accounting, order entry, data processing, telecommunications, security, facilities, and so forth. Each established itself as an individual fiefdom.

The Data Processing Manager position evolved from a numbers-oriented clerk in accounting into one with more technological and programming expertise. It then progressed into a position requiring more than a general overview of business and management disciplines. This has spawned an even more specialized group of managers, with such appropriate titles as Data Processing Manager, Data Base Administrator, Information Systems Manager and Management Information Systems Manager.

At the same time, telephone departments started with technicians having analog and mechanical experience -- primarily because of the less "high tech" automation of the telephone industry and the ever-present AT&T to solve problems. The recent digital evolution along with the divestiture of AT&T is forcing business to make more choices about their telephone systems. It has increased the importance of those experts with advanced telecommunications technical and administrative management experience. The positions are becoming embedded in corporate America's structure, with titles now extending up to Director of Communications.

Telecommunications and data processing started to overlap over the past 20 years. Since 1968, when the Carterphone decision enabled data to be sent legally over telephone lines, data communications has become a major component of data processing budgets. Data communications management became another offshoot of technology, evolving naturally from its integration.

By 1982 it had become apparent that Private Branch Exchange (PBX) telephone systems not only could, but soon would be a major means of information transfer. It was easy to recognize that this continual dispersal of responsibilities could not continue. Parochial interests in the various disciplines then began lobbying and speculating on *who* would control this powerful and flexible mega-technology resource.

Nearly six years later, the jury is still out on who will have control. Some data processing managers and some telecommunications managers are gaining control of the new communications facilities, generally called "Information Systems."

Career Opportunity For Intelligent Building Managers

Within the facilities management field, similar growth toward centralized control over distributed services has taken place. Beginning with a thermostat invented by Warren Johnson in 1886, building automation started with controls dispersed throughout a building or campus. Control once implied an engineer or watchman walking around the building, checking, adjusting, and then bringing all of that information to an office.

Relationships grew between HVAC, electric power, lighting, fire and security control. In the 60's, computers were added for energy optimization and badge access. This was extended from simple computational duties to basic alarming and monitoring systems in the 70's. In addition to centralized monitoring of temperatures and equipment status, energy management functions were developed, followed by fire, security and maintenance management systems.

Those in the intelligent building industry are in the position to fully exploit the advantages of these systems as their capabilities converge. Those who design, build and manage today's intelligent buildings have the broad view necessary to see which information systems should be integrated and which technologies should be used.

Today, facilities management, data processing and telecommunications technology is now being integrated in the nation's "most intelligent" buildings. This integration may range from simple shared reporting between individual systems to a complex central computer controlling all information systems in a multi-building campus. The degree of integration should match the specific strategy guiding the intelligent building's tenants toward their corporate goals. As described in many of the previous chapters, "common sense" determines the degree of integration that is practical.

Just as technology should be integrated where there is a clear, synergistic and strategic result, so too management should be integrated where it makes sense, such as under a CIO. Some managers in the intelligent building industry even have all the skills needed to be CIO's. However, a CIO should be an administrator and a manager first and a technologist second. Early indications are that when one of the "star technical managers" of any individual discipline is promoted to manage the information network, his or her technical background actually inhibits the ability to fairly integrate all the other disciplines. The CIO must be able to diplomatically draw upon a group of advising tacticians who are specialists in their own areas -- the data processing, data communications, telecommunications, facilities and factory communications managers.

Unfortunately, the intelligent building industry is not currently positioned on the inside track of the CIO's evolutionary path. Businesses still look to their data processing and telecommunications staffs if and when they develop strategies for information processing. The intelligent building industry is currently the odd man out of the process.

However, the relative immaturity of the CIO concept offers those in the intelligent building industry a chance to suggest that the CIO scope of authority be expanded. There is still time. The integration of managers of intelligent building technology remains a possibility. Those who design, build and operate the buildings need to emphasize their unique positions in the creation and management of information systems. Their interaction with other managers is what will affect white collar productivity in the next century. And increased productivity is the bottom line which justifies creating intelligent buildings in the first place.

More Information From Buildings Means Better Business

Hopefully, the day is not far off when a corporate officer, overseeing the improvement of profits, will look at timely productivity statistics from the factory, retrieve up-to-the-minute efficiency reports from the building automation systems, use online statistics modules to examine cost-benefit ratios of the data processing operation, and use actual call traffic statistics to optimize the quality of the phone network. It won't be simple, however, if each department continues to string its own wires throughout the building and purchase equipment in a vacuum. Someone has to have a good idea of the overall picture.

The Chief Information Officer is on the horizon. But business is not yet ready; it has not been pushed; business has not experienced any calamities; and the president has not said, "Where's my CIO?"

They are coming. Where there is now redundancy and in-compatibility in information systems and the intelligent build-ing environments in which they operate, there will be produc-tivity, efficiency and flexibility under a CIO. If there are to be CIOs, let's give them control over all of the "information." Sig-nals that are sent along wires to control building environments and TV pictures of deserted corners of warehouse facilities are just as surely information as are voice and data. So too, is what goes on in the factory. Let the CIO oversee everything, from facilities management systems to data processing.

The concerning aspect for the intelligent building industry is that *we* are not thinking more seriously about the concept. Many of us do not see our "data" in the same light as the "infor-mation" that runs a corporation.

The intelligent building industry must let its voice be heard just as data processing and telecommunications have already done. By so doing, those who have the authority to appoint CIOs will realize that the question is broader than "DP or Telecom?"

All the managers of intelligent building technology must work together. That message must be delivered. It makes sense for the opinion leaders in our industry -- consultants, editors, and associations -- to also seriously ponder and debate the broader issues involved with information strategies.

The voice of the intelligent building industry must be heard.

Chapter 30

A Day In The Life Of Tomorrow's Intelligent Building

Robert Heller
Johnson Controls, Inc.

Justin Morgan, president of Morgan Consulting, Inc., is ready to call it day. As he closes the door of his luxurious downtown office space, personnel sensors mounted just inside his office door and in the hallway detect his presence and send messages over the Building Information Transportation System to various building information sub-systems. They respond instantly:

Inside Morgan Consulting's office space, the Helper System in Morgan's office makes a mental note of the time of Justin's departure. The information is sent to the Attendance System for storage. At the same time, the Helper System orders a sensory scan of the office. It finds no other personnel present. Since Justin's departure falls into a time period normally considered an "office unoccupied" time, his Helper System sends a request to the building-wide Helper System. It asks the building-wide Helper to send a message to it if Justin leaves the building. This message, or the failure of Justin to return to the office space in 15 minutes, will generate a message from Justin's Helper to the Comfort and Lighting Control Systems, to put these systems into their unoccupied state.

Meanwhile, the building-wide Helper is setting other events in motion. Having sensed Justin's presence in the hallway, as soon as he appeared, the Helper checked it's database. There was no need to alert security. Justin's presence is a normal occurrence in building off-hours. A message is sent by the building-wide Helper to the building transportation system. A vertical transportation unit is dispatched to Justin's floor. Justin arrives at the transportation area, to be greeted by the arrival of a unit. The doors open and Justin enters. He presses a button for his floor request and also presses the ground transportation request button. A biometric scan of his finger tip identifies Justin to the Transportation Management system and the Ground Transportation Storage System. His floor selection is a valid request, so the vertical building transportation unit speeds off. Meanwhile, the ground transportation storage system "picks" Justin's Excel 7000 unit from the garage inventory and sends off to the Transportation Pad.

Justin arrives at the ground level transportation pad to find his Excel 7000 waiting. He drives off into the night knowing that all environmental systems will shut down automatically, and that the offices of Morgan Consulting are safe and secure.

Science fiction? Or a glimpse into the future?

No one really knows. All we can say at this point is that Justin Morgan's intelligent building will, in all likelihood, be technologically possible, and could become economically feasible sooner than we might think.

Of course, we all have visions of how tomorrow's intelligent building will function.

But as the preceding chapters demonstrate, we're a long way from agreeing upon what constitutes intelligence in today's building. So it's really no surprise that disagreement reigns in any discussion of the future.

In fact, for each of the three broad approaches currently receiving the most attention -- shared telecommunications and data services, information system integration, and the evolution of facilities management systems -- scores of variations are offered by observers from system suppliers to design consultants, from owners to developers.

Whose Vision Is Correct?

Ironically, the group that will play the largest role in determining the shape of tomorrow's intelligent building is the one whose opinion isn't often found in print: the Justin Morgans of the world who will utimately be living and working in the environments we create for them.

Naturally, direct participants in the supply chain to the building marketplace will exert considerable influence. The collective decisions of system suppliers, building designers and contractors will continue to be important.

But, supply is only half the equation. The intelligent buildings that thrive will be those that most closely and completely address the needs and desires of their occupants. It will be their collective opinion about which packages of functional benefits or products best help them meet their personal goals and their perceptions of the financial worth of these products that will ultimately shape tomorrow's intelligent building.

The purchasing decision is the ballot of a free economy. If a product's perceived value allows suppliers to deliver the product to market while earning a profit, the package can become a commercial reality.

It's really unnecessary to debate the relative merits of various technological means of delivering functional benefits in tomorrow's intelligent building. It's irrelevant to argue about whether or not the building design and construction process itself as it currently exists will allow the integration of this technology. And it's short-sighted to focus only on today's vendors, products, technology, buildings or the construction process as it currently exists when discussing the building of the future.

The buyers of tomorrow will elicit changes in all these areas -- changes that will pave the way for the delivery of just the right products, at just the right price. This will be as true tomorrow as it has been in the past. It would therefore behoove us all to focus on these people.

The Buyer Of The Future

Is Justin Morgan going to want all the automation we offered him earlier? Will he appreciate an environment in which increasingly intelligent computer systems respond invisibly to his every move, acting on data supplied by seemingly everpresent sensors? Will he want his car to be automatically guided out of its underground retreat?

And if so, how much will he be willing to pay for it?

These specific questions can't be answered with 100% accuracy today. But we can make some broad assumptions about the future of our society and those who will prosper in it. For instance:

- White collar service businesses will continue to flourish, with a growing emphasis on improved productivity.

- We will witness the maturation of a world economy fueled by world-class products and services.

- The goal of enhancing output while maintaining or reducing the resources used will become even more critical to the health of our economy.

Naturally, one could fill volumes with such predictions about general trends, and feel fairly certain of their accuracy. But let's stop here and consider the import these few thoughts could have on tomorrow's intelligent building:

- In a white collar world vitally concerned with productivity, people will probably want to automate as many routine and repetitive transactions as possible. So the intelligent building of the future could very likely make everything from security checks to comfort control entirely automatic.

- In a world economy, we'll need work environments that accommodate the earth's time zones and cultural differences, permitting full and unfettered communication any time of the day or night. This suggests that tomorrow's building will have to extend its capabilities to occupants 24 hours a day, 365 days a year, both on-site and from remote locations.

- In an environment that rewards improving the ratio of output achieved to resources used, people will probably welcome systems that streamline the handling of both routine and non-routine transactions that don't relate to their higher level goals. As a result, intelligent buildings will most likely make increasing use of artificial intelligence to help their occupants achieve speedy resolution of exceptional situations.

Beware Oversimplification

These observations, of course, are gross oversimplifications.

There may be those who question the probability of such changes occurring; they are, after all, different from today's norm. But we need only look to the recent past to see a substantially different state of affairs in a backward direction. Why would it be unlikely that that should be substantially different when looking forward in time?

Consider, for example, the history of HVAC systems in commercial and public buildings.

Not so very long ago, these systems were fairly inflexible.

The occupants had little control over their heating, ventilating and air conditioning services during normal business hours -- and even less during non-conventional working hours.

Not that it mattered. A 9-to-5 world was pretty much the rule.

Not so very long ago, HVAC systems consumed energy with impunity. Building owners and managers had no means of evaluating consumption, other than comparing invoices from their local utilities -- in many ways a futile exercise, since they could do little to control energy use, anyway.

Not that it mattered. In those days, energy was both plentiful and cheap.

Not so very long ago, access control and security were handled via locks, keys and on-site security personnel. Strangers demanding access to a building were processed like many other non-routine transactions -- manually, by asking for identification, requiring sign-in and perhaps confirming their business via a quick call to those they were visiting.

Not that it mattered. People used to stay in one job year after year. Strangers and visitors were the exception, and in most building environments, presented themselves in relatively short and predictable time spans -- specifically, during the 9 to 5 workday.

And so on: Similar forces shaped the way we handled everything from lighting control to fire management.

But all that has changed. Circumstances change resulting in systems adaptation to the new circumstances. Unconventional business hours, skyrocketing energy costs, an increasingly mobile work force, and mixed-use structures have all contributed to creating a far different environment. Today, sophisticated energy, security, lighting and fire management systems have become quite common; indeed, the building without such systems may be well on the way to extinction.

It's clear that, whatever shape tomorrow's intelligent building may take, it will be the result of an elaborate network of forces with one thing in common: the nature of the building's owners and occupants.

What Lies Ahead?

Naturally, those suppliers who prosper in tomorrow's building marketplace will be those who listen carefully and successfully anticipate the proper timing and pricing of products that meet owner and occupant needs. A wait-and-see attitude is not very practical in this rapidly evolving market.

My prediction: the intelligent building of the future will feature products that streamline the processing of both routine and exceptional transactions. Its systems will feature enhanced communications bi-directionally between the user and the computer, as well as among computers themselves.

Consider in today's market, for instance, the growing use of Touch Tone telephones for extending communications with a facilities management system to building occupants. It's already possible for an occupant to excercise some control over the systems in his building environment by picking up the phone and punching in a series of numbers -- numbers which are sent to the facility's management system as a control request. And voice recognition technology is emerging to make this interaction even simpler and more "natural."

We will most likely see a steady growth in the application of artificial intelligence and expert systems to help computers "think through" the handling of exceptions, thereby minimizing the need for input from experienced operators.

Advances in sensor technology, biometrics, and overall communications capabilities will also shape our buildings' futures. We may soon be able to undergo a security check simply by talking to a computer capable of analyzing our speech patterns...and expect a chain of computer-to-computer communications to ensue.

How much of this will become reality? How soon?

That will depend upon a broad range of contributing factors.

And these factors will continue to change rapidly.

As a society, we once resisted depersonalization in our everyday lives. Many of us preferred standing in line at the bank, for instance, to managing our finances via automatic tellers. The growth in electronic funds transfer networks and terminals demonstrates that our priorities have shifted enough to accommodate this degree of depersonalization.

Similarly, conventional wisdom says that we would balk at talking to computers, and at listening to their responses...that we would view as "big brotherish" a biometric scanning system capable of permitting or denying access to a building without any need for human intervention in the "decision" process.

Yet social and political forces are changing even these attitudes. The concept of a biometric scanner, for instance, is likely to gain increasing acceptance if for no other reason than the rise in crime and terrorism worldwide.

Predicting A Day In The Life

That's as far as I'll go in predicting which technological candidates will win out in the race to control tomorrow's intelligent building.

For those who demand certainty before making decisions, the only solution is to wait until tomorrow comes.

For those of us who can't afford to do so, the only alternative is to go out into the marketplace to talk with -- or, more accurately, to listen to -- tomorrow's buyers as they discuss their own needs and desires.

APPENDIX A

The U.S. intelligent buildings industry has suffered from a lack of identity -- so many different types of buildings have been labeled intelligent by their owners during the past four years, that the term "intelligent building" ran the risk of losing its meaning.

The Intelligent Buildings Institute (IBI) representing most major participants in the intelligent building community, has moved to correct the identity crisis. It has published the Intelligent Building Definition, an easy-to-read handbook which captures the essential elements of intelligent building design.

"The building community absolutely needed to define what it was building and buying," said Ronald J. Caffrey, chairman of IBI's Board of Directors. Caffrey, vice president of Johnson Controls, a leading provider of intelligent building control devices, continued:

"There has been too much confusion for both vendors and end users. It was time the intelligent building community reached agreement on just what constituted a so-called 'intelligent building'."

The book was developed by IBI committees composed of nationally-recognized leaders including Honeywell Incorporated, RealCom Communications Corporation (an IBM Company), Tishman Research, Syska & Hennessy, Carlini and Associates, Bell of Pennsylvania and other domestic and international organizations.

"The definition is important," said Theodore Schell, president of RealCom Communications and a member of IBI's Board of Directors. "...it is a concept which needed to be explained ...it doesn't say how every individual building will be constructed, but provides a conceptual framework of how intelligent buildings should be perceived."

"IBI has given the real estate market a new outlook on the future," according to James Carlini, president of Carlini and Associates. Carlini served as chairman of the subcommittee which developed the initial draft of the Definition handbook. "It is the first effort of a broad cross section of companies to crystalize the meaning of an intelligent building, and to define terminologies used in the industry," he said.

The term "intelligent building" is used to describe a building capable of achieving optimum life cycle costs of occupancy and increased organizational productivity through its inherent design and management. The handbook defines an intelligent building as "...one which provides a productive and cost-effective environment through optimization of its basic four elements -- structure, systems, services and management -- and the interrelationships between them."

These interrelationships are explained in greater detail in the later sections. More than 140 commonly used terms are defined, and a "building suitability analysis" is provided.

"The Institute is offering more than a booklet," said Richard Geissler, IBI executive director. "For the initial price of $25 purchasers will receive the Definition booklet, a binder for future supplements, and an opportunity to register for upcoming revisions. Defining intelligent buildings and identifying their systems and subsystems is a continuing process; to keep abreast of industry developments IBI will be changing the book to incorporate new and innovative material."

IBI is a not-for-profit professional association established in January 1986 to serve the widest possible constituency in all phases of intelligent buildings. Among its programs are technical efforts to define specific terms in intelligent buildings; government relations; development of open protocol for automated building communications systems; market development; and training programs.

The Intelligent Building Definition handbook can be ordered from IBI, 2101 L Street, N.W., Washington, DC 20037, (202) 457-1988.

GLOSSARY

Abbreviated Dialing

Enables a caller to dial a frequently used number by means of fewer digits than the entire telephone number. Also called speed dialing.

Acceptance Test

Operating and testing of a new system to ensure that the system is operating satisfactorily before being accepted.

Access Control

Means of gaining access to a network (i.e., CSMA/CD, Token Passing, Polling).

Access Control System

Computerized building security equipment such as badge readers for entry protection.

Access Line

Connection from the customer to the local telephone company for access to the telephone network; known as local loop.

Actuator

A device that converts a pneumatic or electric signal to a force which produces movement. Example: air damper.

Acoustic Coupler

Device which allows a telephone to be used for access to the switched telephone network for data transmission where digital signals are modulated as sound waves. Sometimes called a modem.

Adapter

A device for making connections between a cable and other cables or equipment that would otherwise be incompatible.

Adaptive Differential Pulse Code Modulation (ADPCM)

Encoding technique (CCITT) that allows analog voice signals to be carried on a 32K bps digital channel. Sampling is done at 8KHz with 3 or 4 bits used to describe the difference between adjacent samples.

Adaptive Routing

Routing that automatically adjusts to network changes like traffic patterns or failures.

Adjunct Power

Power supplied to optional data or voice equipment, such as speakerphones, by power supplies located in an equipment room, satellite closet, or user room.

Administration Point

A location at which communication circuits are administered, i.e., rearranged or rerouted, by means of cross connections, interconnections, or information outlets.

Administration Subsystem

That part of a premises distribution system where circuits can be rearranged or rerouted. It includes cross-connect hardware, interconnect hardware, and jacks used as information outlets.

AE

Commonly used acronym for application engineer or architect engineer. AE's are often responsible for design of a building automation system.

AHU

Air handling unit. A unit consists of a fan which moves air through dampers and ducts over hot and/or cold coils and into one or more rooms. Traditional technique for heating, ventilating and air conditioning commercial buildings. Fan types can be single or variable speeds. Controls may be digital, electric or pneumatic.

Alarm Display

Indicators on attendant telephone console that show the status of the system. Usually two alarms are included: a minor alarm and a major alarm.

ALLIANCE®

Johnson Controls unique family of flexible building service programs that provide equipment, maintenance, repair and trained personnel to keep commercial buildings operating comfortably.

Alternative Routing

Method of routing whereby a different communications path is used if the usual one is not available.

American Standard Code for Information Interchanged (ASCII)

Eight-bit code yielding 128 characters, both displayed and non-displayed (for device control); used for text transmission.

Amplitude Modulation

A method of modifying a sine wave where amplitude is modified in accordance with the information to be transmitted.

Analog Point

A piece of equipment or a sensor which has a continuous range of settings or values that can be monitored or controlled by a building automation system. Examples: damper, temperature sensor.

Analog Signal

Signal in the form of a continuously varying quantity such as voltage, which reflects variations, such as loudness.

Analog Transmission

The transmission of a continuously variable signal as opposed to a discretely variable signal.

Annunciator Command Terminal

Fireman's control panel typically located in the lobby so the fire department can operate a building automation system during an emergency.

Artificial Intelligence

Computer systems that solve problems and operate symbolically rather than algorithmically. The theory implies an emulation of human intelligence by computers.

ASCII

American Standard Code for Information Interchange. (An accepted standard for computer data transmission).

ASR

Auto-speech recognition.

ATC

Automatic Temperature Control. Electro-mechanical control equipment. Typically not integrated into building automation systems because they are not digital.

Attendant

Usually refers to a local switchboard operator.

Attendant Console

Centralized operator console, either desktop or floor-mounted,that uses pushbutton keys for all control and functions.

Attenuation

Decrease in magnitude of the current, voltage, or power of a signal in transmission between points because of the transmission medium. Expressed in decibels (Db).

Audio Frequencies

Frequencies that correspond to those heard by the human ear (usually 30Hz to 18,000 Hz).

Automated Attendant System

Computer controlled system that performs most attendant functions, such as answering calls, extending them to station users, taking messages, and providing assistance. The system provides voice prompts, to which users reply via any standard 12-button touchtone dialpad.

Automatic dialer

Device that allows the user to dial preprogrammed numbers by pushing a single button. Also, an auto dialer.

Automatic Route Selection (ARS)

Provides automatic routing of outgoing calls over alternative customer facilities based on the dialed long distance number.

The station user dials either a network access code or a special ARS access code followed by the number. The PABX routes the call over the first available special trunk facility, checking in a customer-specified order. Alternative routes can also include tie trunks to a distant PABX. When such routing is used, the restriction level associated with the call can be transmitted to the distant PABX as a travelling class mark. Incoming tie trunks from other locations (e.g., main or satellite) can be arranged to have automatic access to ARS. This allows station users at these systems to dial a single access code to use the ARS feature at a distant PABX.

AVD Circuits

Alternate Voice/Data circuits that have been conditioned to handle both voice and data.

Backbone Closet

A room where backbone cables are terminated and where cross-connect hardware is located for circuit administration. Sometimes called a "riser closet."

Backbone Subsystem

The section of a premises distribution system that includes the cabling (riser cable) that runs from an equipment room to the various floors. In a building on a single floor, the backbone subsystem is the trunk or main channel of the communications system.

Band

1) A range of frequencies between two limits. 2) Regarding WATS service, the specific geographical area where a customer is entitled to call.

Bandwidth

A range of frequencies expressed in Hertz (cycles per second). If a system or medium can accommodate signals in the frequency range from 40,000 Hertz to 70,000 Hertz, its bandwidth would be 30,000 Hertz (30 kiloHertz). The greater the bandwidth, the more aggregate information can be conveyed.

BAS

Building Automation System. (See below)

Baseband

High-speed communications scheme usually in coax where a signal uses the entire bandwidth for its transmission.

Baseband Signaling

Transmission of a digital or analog signal at its original frequencies; not changed by modulation.

Basic Rate Interface

In ISDN, the interface to the basic rate CCITT 2B + D, 2 channels + 1 signaling channel.

Baud

A unit of data transmission speed equal to one bit per second.

Communications speed between computers, field devices, and input/output devices is usually described using baud rates.

BCD

Binary Coded Decimal, a 6-bit alphanumeric code.

BCN

Building Communications Network. A network which integrates, to the greatest possible extent, telecommunications, data processing and building automation communications transmission over a united wiring plan.

Bell Operating Company (BOC)

Any of the divested operating telephone companies in the United States, such as Wisconsin Bell, New Jersey Bell, and Mountain Bell.

Binary Code

Representation of numbers expressed in the base-2 number system.

Binary Point

A piece of equipment or a sensor which has only two possible contact states that can be monitored or controlled by a building automation system. Examples: single-speed fan, door sensor.

Bipolar

Type of integrated circuit that uses both positively and negatively charged currents, characterized by high speed and cost.

Bisync

Binary Synchronous Communications (BSC). Character-oriented data communications protocol developed by IBM; oriented towards half-duplex link operation.

Bit

Contraction of "binary digit," the smallest unit of information in a binary system.

Bit Rate

Speed at which bits are transmitted, usually expressed in bits per second.

Bits Per Second

Basic unit of measurement for serial data transmission capacity; abbreviated as K bps, or kilobit/s, for thousands of bits per second; M bps, or megabit/s, for millions of bits per second; G bps, or gigabit/s for billions of bits per second.

Blocking

Inability to interconnect two lines in a telephone network because all possible paths between them are already in use.

Broadband

High-speed communications scheme where the very high bandwidth is divided into smaller, discrete components. Many different signals can be transmitted simultaneously.

Building Automation System (BAS)

A modular array of compatible computer components that automate a wide-range of building operations from heating, ventilating and air conditioning to energy management, maintenance management, fire management, security and lighting control.

Bus

Horizontal and vertical bars which distribute line-side electric power to the components of a motor control center.

Also used in distributed computer networks to describe the interconnecting wiring planes.

Byte

Small group of bits of data that is handled as a unit. Usually is an 8-bit byte.

CAL1 (Control Application Language)

High-level programming language which allows the user to build

DSC-8500^{TM} programs. This language includes direct digital control functions. It is customized to simplify writing programs that control heating, ventilating and air conditioning equipment.

Call Accounting System

A computer system that tracks outgoing calls and records data for reporting.

Calling Party

The person who originates a call. Also known as calling subscriber.

Call Processing

Sequence of operations performed by a switch from the acceptance of an incoming call through its final disposition.

Call Restriction

PBX feature that prevents selected extension stations from dialing external calls or reaching the operator except through the PBX attendant.

Campus Subsystem

The section of a premises distribution system that includes the cabling or other transmission media for communications circuits between or among buildings within a campus or office park.

Carrier (CXR)

1) Continuous frequency capable of being modulated with a second signal. 2) Communications company or authority providing circuits to carry traffic (also known as common carrier firms other than AT&T are other common carriers (OCC's).

Carrier Frequency

Frequency of the carrier wave modulated to transmit signals.

Carrier Signaling

Any mode of the signaling techniques used in multichannel carrier transmission. The most commonly used techniques are in-band signaling, out-of-band signaling, and separate channel signaling.

Carterfone Decision

The landmark 1968 FCC decision that first permitted the electrical connection of customer-owned terminal equipment to the telephone network.

Cathode Ray Tube (CRT)

"Television screen" or "video display terminal" used to display information from a building automation system or computer.

CATV

Community Antenna Television. Used for cable television within a building or complex. Can be integrated into building communications network.

CB

Composite Triple Beat Distortion

CBX

1) Centralized Branch Exchange. 2) Computerized Branch Exchange.

CCITT

Comite Consultatif International de Telephonie et de Telegraphie. An advisory committee to the International Telecommunications Union (ITU) whose recommendations and standards covering telephony and telegraphy have international influence among telecommunications professionals.

CCTV

Closed Circuit Television. Used for building security. Can be integrated into a building automation system.

CDR

Call Detail Recording.

CE

Commonly used acronym for consulting engineer or specifying engineer. CE's are often responsible for design of a building automation system.

Central Office (CO)

The main load facility for the local exchange communication switch.

Central Processing Unit (CPU)

Computer controlling the interpretation and execution of instructions.

Centrex

Service that allows every subscriber to be directly dialed from the outside. Centrex switching equipment is located in the central office and allows subscribers to access facilities normally provided by a separate PBX.

Change-of-State (COS)

When a binary or analog point experiences a change from Normal to Abnormal condition or a return from Abnormal to Normal condition. One key feature of a building automation system is the ability to report change-of-states in a realtime environment.

Channel

A transmission path of signals between two or more points. Also called line, circuit, link, TIC path, or facility.

Chiller

Large commercial mechanical equipment that cools water for air conditioning. Equipment can use either centrifugal compressor or steam absorption process to cool water. (See Ton for capacity rating)

Chiller Manager

Energy-saving software program that controls chilled water reset temperatures, condenser water reset temperatures and sequences chillers, if there is more than one, to improve equipment performance based on actual chilled water demand. Chillers are typically the largest single energy consumer in a building.

Chip

Substrate upon which LSI or VLSI circuits are fabricated. Sometimes used to refer to the circuits and substrates themselves.

Circuit

1) A means of two-way communication between two or more points.
2) A group of electrical/electronic components connected together to perform a function.

Circuit Switching

A temporary direct connection of one or more channels between two or more points to provide the user with exclusive use of an open channel. A discrete circuit path is set up between the incoming and outgoing lines. Also called line switching.

CIU

Communications Interface Unit.

Class of Service

Categorization of telephone users according to specific type of telephone use. Also applies to PBX station user features and restrictions.

Clipping

Loss parts of words due to operation of voice-actuated devices.

Closed Circuit Television (CCTV)

Television transmission via direct link between two points, as opposed to broadcast transmission to many points.

Closed Loop System

Connection of heating, ventilating and air conditioning components to allow system feedback (e.g., heating unit, valve and thermostat arranged so that each component affects the other and can react to it.)

CMOS

Complementary Metal-Oxide Semiconductor (logic circuit).

C/N

Carrier to Noise Ratio.

CO

Central Office.

Coaxial Cable

A cable with a least one transmission line consisting of two conductors, an inner conductor and an outer conductor, insulated from one another by a dielectric. Coaxial cable carries higher frequencies than twisted pair cable and offers a broader signal bandwidth. It is commonly used to transmit video signals, but can also be used for certain high-speed data applications.

Codec

Coder-decoder device used to convert analog signals, such as speech or video, to digital form for transmission over digital media, and back again to the original analog form. A coder is required at each end of the channel.

Cogeneration

Using electric power generation equipment for buildings. Typically used to supplement public utility service during peak rate periods. Can be integrated into building automation system.

Coin Box

A telephone requiring insertion of coins or credit cards before it can be used.

Common Carrier

A business providing regulated telephone, telegraph, telex, and data communications services.

Communications Satellite

Earth satellite designed to act as a telecommunications relay. Usually in geosynchronous orbit 35,800 kilometers above the equator so it appears to be stationary in space.

Concentrator

A device that connects a number of circuits that are not all used at once to a smaller group of circuits for economical transmission. A telephone concentrator achieves the reduction with a circuit-switching mechanism.

Conditioning

Procedure to make transmission impairments of a circuit lie within certain specified limits and tolerances, typically used on telephone lines leased for data transmission.

Conductor

Any wire or cable that can carry an electric current.

Conduit

A pipe, made of rigid material, that offers continuous support and protection for cables. Conduit may be used in both the backbone (riser) and horizontal wiring subsystems. When backbone (riser) closets are not aligned, conduit is used to provide channels for pulling cable from floor to floor. In the horizontal wiring subsystem, conduit may be used between a backbone or satellite closet and an information outlet in an office or other room.

Conference Call

Call established among three or more stations in such a manner that each of the stations is able to carry on a communication with the others.

Control Character

Character that initiates, changes, or stops a control operation.

Controller

Pneumatic, electronic, or digital device which determines and regulates the position of controlled devices such as valves, dampers and contracts based upon external inputs such as temperature.

Control Signals

Signals that pass from one part of a communications system to another to control the system.

CPE

Customer Premises Equipment. Terminal equipment connected to the telephone network.

CPI

Computer to PBX Interface. Gateway providing 24 digital PCM channels at an aggregate rate of 1.544M bps. Developed by Northern Telecom. See also DMI.

Cross Connection

A circuit administration point using one or more jumper wires or a patch cord for circuit rearrangements. Typically, cross connections are located in an equipment room or in a wiring closet.

Crosstalk

Unwanted transfer of signal from one circuit to another circuit.

CSMA/CD

Carrier Sense Multiple Access with Collision Detections.

Cutover

Changing of lines physically from one system to another, usually done at the time of a new system installation.

Cybertronic®

Johnson Controls electronic control equipment. Cybertronic equipment is principally proportional control with resistance sensing elements, whereas electrical equipment is principally two-position control with contacts opened and closed by the sensing elements.

Data Base

Portion of software which consists of lists and files of information. In a building automation system, this primarily consists of binary and analog point names and descriptions such as temperature limits. It also includes operating information such as passwords and programs.

Data Circuit

A communications facility that allows transmission of data in either direction, in analog or digital form.

Data Line Interface (DLI)

Point at which a data line is connected to the telephone system.

Data Link

Any serial data communications transmission path, between two adjacent nodes or devices, usually without any intermediate switching nodes.

Data Terminal Equipment (DTE)

Generally end-user devices, such as terminals and computers, which either generate or receive the data carried by the network.

Data Transmission

Process of transmitting digitally-ended information. As more PABX systems accept digital data as well as voice signals for switching and transmission, they allow a user with a central computer and remote terminals to provide dial-up service to connect the terminals with the computer without the use of modems.

dB

Decibel. A unit for measuring relative strength of a signal parameter. The number of decibels is ten times the logarithm (base 10) of the ratio of the power of two signals, or ratio of the power of one signal to a reference level.

DDC

Direct Department Calling.

DDD

Direct Distance Dialing.

DDI

Direct Dialing In. See Direct Inward Dialing.

Dedicated

Used exclusively for a single purpose or subscriber.

Dedicated Circuit

A communications link used exclusively by one person at each end.

Demand Limiting

The process of keeping electrical consumption at or below a specified target level by switching off specified loads (e.g., equipment such as chillers, fans, lights). The target level(s) is determined by electric utilities which typically bill commercial customers based on peak usage as well as kWh. (Also referred to as Peak Shaving or Load Shedding.)

Demarc

The <u>Demarc</u>ation point between carrier equipment and customer-premises equipment (CPE) which is usually a terminal block.

Dial Tone (DT)

A signal sent to an operator or subscriber indicating that the switch is connected.

Dielectric

A substance that does not conduct electrical current.

Digital

Referring to communications procedures, techniques, and equipment whereby information is encoded as either binary "1" or "0"; as opposed to the analog representation of information in variable, continuous, waveforms.

Digital Network

A network incorporating both digital switching and digital transmission.

Digital System Controller (DSC)

Digital computer which periodically compares analog and binary point status to the desired setpoint or state and can directly initiate control action if necessary. Also referred to as Direct Digital Control (DDC). Digital system controllers may be integrated into a building automation system.

Direct Digital Control (DDC)

Refer to Digital System Controller.

Direct Distance Dialing (DDD)

Telephone exchange service that enables the phone user to call long distances without operator assistance.

Direct-in Lines

Allow direct termination of exchange lines to station instruments, bypassing the attendant console.

Direct Inward Dialing (DID)

Incoming calls from the exchange network which can be completed to specific station lines without attendant assistance. Also called Direct Dialing In (DDI).

Direct Outward Dialing (DOD)

Allows a PABX, Centrex, or hybrid system user to gain access to the exchange network without attendant assistance.

Disk

Rotating magnetic storage device.

Dispersion

The tendency of a beam of light to spread out and lose its focus.

Distribution Frame

A structure (typically wall-mounted) for terminating telephone wiring, where cross-connections are made to extensions. Also called distribution block.

Divestiture

The break-up of AT&T in 1984 that included the organization of 22 AT&T-owned local Bell Operating Companies (BOCs) into seven independent Regional Bell Holding Companies (RBHCs).

DMI

Digital Multiplexed Interface. A gateway providing 23 digital PCM channels + 1 signaling channel at an aggregate rate of 1.544Mbps. Developed by AT&T.

Downtime

The total time a system is out of service due to failure.

DPPM

Differential Pulse-position Modulation.

DSC-1000^{TM}

Johnson Controls family of microprocessor-based digital control products which are preprogrammed for specific heating, ventilating and air conditioning control functions. Can be integrated into a Johnson Controls DSC-8500 network.

DSC-8500^{TM}

Johnson Controls digital system controller programmable for heating, ventilating and air conditioning control functions as well as other applications using CAL1. Can be integrated into a Johnson Controls JC/85 building automation system.

DTE

Data Terminal Equipment.

DTMF

Dual Tone Multifrequency signaling. The basis for operation of pushbutton telephone sets in which a matrix combination of two frequencies, each from groups of four, are used to transmit numerical address information. the two groups of four frequencies are 697 Hz, 770 Hz, 852 Hz, and 941 Hz, and 1209 Hz, 1336 Hz, 1477 Hz, and 1633 Hz.

Duplex Circuit

A circuit used for transmission in both directions as the same time.

Duty Cycling

(Also referred to as load cycling or load rolling.) The process of reducing total electrical consumption by intermittently stopping equipment.

Dynamic

Any information that will change automatically when displayed on a building automation system CRT or display panel. If the status of building equipment changes, such as a temperature displayed on a CRT, the value is updated when the actual value changes.

Earth Station

Ground-based equipment needed to communicate via satellites. Also called ground station.

EBCDIC

Extended Binary-Coded Decimal Interchange. Code: an 8-bit alphanumeric code used primarily by IBM.

EIA

Electronics Institute of America.

EIA Interface

A standardized set of signal characteristics (time, voltage, and current) specified by the Electronic Industries Association.

Electromagnetic Interference (EMI)

Radiation leakage outside a transmission medium which results from the use of high-frequency signal modulation. EMI can be reduced by appropriate shielding.

Elevator Recall

Overriding control of normal elevator operation by the fire management system. Elevators are automatically directed to ground floor the instant an alarm is sound, and placed under full control of the Fire Department.

EMI

Electromagnetic Interference.

Energy Management System (EMS)

Computerized monitoring equipment for managing energy use. Usually related to HVAC and lighting systems. May include limited control functions.

Enthalpy

Total heat content measured in Btus per pound (or kJ/kg).

This measurement takes into account the latent heat of humidity in the air as well as the measurable temperature itself. To save energy, a building automation system can take into account the latent heat of humidity as it conditions air.

EOL

End Of Line Termination.

Equal Access

Department of Justic ruling (effective 9/84) that requires all RBHCs with ESS systems using SPC technology, and serving a market of at least 10,000 access lines, to offer the same quality of connection at the same rates to all common carriers. All called dial 1 access.

Equipment Room

An enclosed space in which voice and data common equipment is located and where circuit administration is performed.

Equipment Wiring Subsystem

The section of a premises distribution system that includes the cables and connectors between units of equipment in an equipment room.

Exchange

Assembly of equipment in a communications system that controls the connection of incoming and outgoing lines, including necessary signaling and supervisory functions.

Exchange, Private Automatic Branch (PABX)

Private automatic phone exchange that provides for the switching of calls with internally and external telephone network.

Exchange, Private Branch (PBX)

Private, manually operated telephone exchange that provides private telephone service within an organization and allows access to the public telephone network.

Exchange Area

Area containing subscribers served by an exchange.

Extended Area Service (EAS)

Option where the telephone subscriber can pay a higher flat rate to obtain wider geographical coverage without additional message unit charges.

Extension Telephone

An additional telephone set on the same line but at a different location.

Facsimile

A system for the transmission of images. The image is scanned at the transmitter and reconstructed and printed at the receiving station.

Fail Soft

When a piece of equipment fails, lets the system fall back to a degraded mode of operation rather than it failing completely.

FCC

U.S. Federal Communications Commission. A Washington, DC-based regulatory agency established by the Communications Act of 1934, charged with regulating all electrical and radio communications in the U.S.

FH

Frequency Hopped.

Fiber Optics

The transmission of information as light pulses along fiber optic waveguides.

Fiber Optic Waveguides

Thin filaments of glass through which a light beam can be transmitted by means of multiple internal reflections. Other transparent materials, such as plastic, may also be used.

FID

Field Interface Device.

Firmware

Computer programs or instructions that are stored in the Read Only Memory (ROM). It is analogous to software, except it is in hardware form. Firmware is loaded into the equipment either at the time it is manufactured or later for preprogrammed controllers such as the DSC-1000™.

Flat Rate

A fixed payment for service within a geographic area.

Foreign Exchange (FX)

(U.S. term) Connects a customer's location to a remote exchange. This service provides the equivalent of local service from a distant exchange.

Frequency Modulation (FM)

One of three ways of modifying a sine wave signal to make it carry information. The sine wave has its frequency modified in accordance with the information to be transmitted.

Full Duplex (FDX)

A communications system capable of transmission simultaneously in two directions.

FX

Foreign Exchange.

Gain

An increase in signal power in transmission from one point to another, expressed in dB.

Grade of Service

Quality of telephone service provided by a system described in terms of the probability that a call will encounter a busy signal during the busiest hour of use.

Half Duplex (HD or HDX)

A communications system capable of transmission in either direction, but not in both directions simultaneously.

Hardcopy

Printed output of a computer. Example: Reports.

Hardware

Physical equipment associated with a computer system. Includes the Central Processing Unit (CPU), CRT, disk drive, controller(s) and analog or binary point modules. Field hardware includes digital equipment such as controls, sensors and actuators.

Hard Wired

A communications link that permanently connects two nodes, stations, or devices.

HDLC

High-level Data Link Control. Bit-oriented communication protocol developed by the ISO.

Hertz (Hz)

Unit of frequency equal to one cycle per second.

HF

High Frequency.

Historical Device

Magnetic storage file in computer memory where reports and other records may be accumulated until requested by the operator for display on a CRT or printing.

Hookswitch

See Switchhook.

Hoot and Holler Circuit

Four-wire, private line circuit using voice signaling or ringdown to transmit voice information among dispersed sites, such as branch offices. Used primarily in financial institutions to provide timely communication of investment information among traders.

Horizontal Wiring Subsystem

The section of a premises distribution system that includes the cables running between a backbone or satellite closet and an information outlet. Sometimes called gray cable.

Hot Line

Line serving two telephone sets exclusively, on which one set will ring immediately after the receiver of the other set is lifted.

HVAC

Heating, Ventilating and Air Conditioning.

Hz

Hertz.

IDF

Intermediate Distributing Frame. A distributed frame from a communication network.

Information Outlet

Typically the 8-pin modular jack in a user room; a circuit administration point for connecting work location wiring to other premises distribution wiring.

Input/Output Device

Any device used to communicate with a building automation system computer. Examples: Keyboard, Annunciator Command Terminal, Telephone.

Intelligent Buildings

Building "intelligence" exists on a continuum. Johnson Controls began making buildings intelligent when Professor Warren Johnson invented the first electric thermostat and founded the company. Buildings can be advanced on the continuum of "intelligence" by adding central monitoring and control for thermostats and other HVAC devices. "Intelligence" is furthered by installing fire and security management, office automation and telecommunications systems. And it is even further extended by integrating these systems into advanced telecommunications and building automation systems, also by providing sophisticated technology to individuals through shared tenant services.

Intelligent Buildings Institute (IBI)

Trade association formed March, 1986 to promote the intelligent building industry. Non-profit professional organization chartered to provide market information, joint R&D, government advocacy, and regulatory action as well as general education. Chairman is Ronald J. Caffrey, Vice President of Marketing, Johnson Controls, Inc.

Integrated Services Digital Network (ISDN)

Project underway within the CCITT for the standardization of operating parameters and interfaces for a network that will allow a variety of mixed digital transmission services to be accommodated. Access channels under definition include a basic rate defined by CCITT 2B + D (64K + 64K + 16K bps, or 144K bps) and a primary rate that is DS-1 (1.544M bps in the U.S., Japan, and Canada, and 2.048M bps in Europe).

Integrated Voice/Data Terminal (IVDT)

One of a new family of devices that features a terminal keyboard/display and voice telephone instrument. Many contain varying degrees of local processing power; can be designed to work with a specific customer premises PABX, or else be PABX independent.

Intercom

Internal communications system that allows calling generally within the same location, but not outside of the system.

Interconnect Company

An organization, other than the telephone company, that supplies telephone equipment and maintenance services by sale, rental, or leasing.

Interconnection

A circuit administration point, other than an information outlet or cross connection, that provides limited rearrangement capability without patch cords or jumper wires. Typically, it is the modular jack and plug arrangement used for circuit administration in small distribution systems.

Interlocking

Software program that allows the user to have a particular event or set of events trigger another event or sequence of events. Example: If a temperature exceeds its high limit then this causes a fan to be started.

Intermediate Distributing Frame (IDF)

Frame having distributing blocks on both sides, permitting connection of any telephone line with any line circuit.

International Standards Organization (ISO)

Agency of the United Nations concerned with international standardization in a broad range of technical fields.

International Telecommunication Union (ITU)

Telecommunications agency of the United Nations, established to provide standardized communications procedures and practices including world-wide frequency allocation and radio regulations.

ISDN

Integrated Services Digital Network (emerging communications standard).

ISO

International Standards Organization.

IVDT

Integrated Voice/Data Terminal.

Jack

A device used generally for terminating the permanent wiring of a circuit. Usually accessed by the insertion of a plug.

JC/85®

Johnson Controls building automation system. A modular array of compatible computer components that can automate a wide-range of building operations in almost every size building. Each system precisely fits building automation requirements whether they are for a high school or a multi-tower office complex.

JC/Link

Communications interface that enables Johnson Controls JC/85® Building Automation System to operate over a twisted-pair copper wire telephone network. The link is compatible with AT&T, Northern Telecom, ROLM, GTE, InteCom, Mitel and other PBX equipment.

JC/Telephone I/O (JC/TIO)

Telephone input/output device that enables Johnson Controls JC/85® Building Automation System to be operated by dual-tone multi-frequency (e.g., Touchtone® trademark of AT&T) input with synthesized voice output.

JC/Workstation

An integrated voice data terminal for Johnson Controls JC/85® Building Automation System.

Jumper Wire

A short length of wire used to link two cross-connect termination points.

K bps aKb

Kilobits per second.

Keyword

Abbreviation (mnemonic) used in building automation systems to allow operators to communicate computer instructions in recognizable commands. Example: STA and STO are the commands for start and stop.

Key Telephone System (KTS)

When more than one telephone line per set is required, key telephone systems offer flexibility and a wide variety of uses, e.g., pickup of several exchange lines, PABX station lines, private lines, and intercommunicating lines through push buttons. Features of the system include pickup and holding, intercommunications, visual and audible signals, cutoff, exclusion, and signaling.

KHz

Kilohertz, one thousand cycles per second.

Ku Band

Portion of the signal spectrum, being used principally for satellite communications. Frequencies approximately in the 12GHz to 14GHz range.

KW

Kilowatt.

KWh

Kilowatt hour.

LAN

Local Area Network.

Laser

Acronym for Light Amplification by Stimulated Emission of Radiation. Lasers convert electrical signals into radiant energy in the visible or infrared parts of the spectrum. They are widely used in fiber optic communications.

LATA

Local Access and Transport Area.

Leased Line

A dedicated circuit that connects two or more user locations and is for the use of the subscriber. Such circuits are generally voice grade analog, used for voice or data, and point-to-point or multipoint. Also called "private line".

Least Cost Routing (LCR)

See Automatic Route Selection (ARS).

LED

Light Emitting Diode.

Line

Transmission path from a nonswitching subscriber device to a switching system.

Line Speed

The data rate that can be transmitted reliably over a line.

Link

The circuit between two points.

Link Level

Layer in OSI Reference Model which transforms transmission provided by physical level into an error-free line.

Load Manager

Energy-saving software program that controls demand limiting, duty cycling and electrical load processing based on actual electrical consumption.

Load Processing

The process of assigning priorities to electrical loads (e.g., minimum and maximum on/off times, temperature overrides).

Local Access and Transport Area (LATA)

Geographic regions within the U.S. that define areas where the Bell Operating Companies (BOC's) can offer exchange and exchange access services (local calls, private lines, etc.).

Local Area Network (LAN)

One of the several types of distance limited communications networks intended for applications such as data transfer, text, facsimile, and video.

Local Call

Any call for a destination within the local service area of the callers station.

Local Exchange

Exchange in which subscribers' lines terminate. The exchange has access to other exchanges and to national trunk networks.

Also called local central office, end office.

Local Loop

The communications circuit between the subscriber's equipment and local exchange's equipment.

Long Distance

Any telephone call to a destination outside the local service area of the caller. Also called "toll call".

Long-Haul

Long distance; telephone circuits that cross out of the local exchange.

Loop

The local circuit between the exchange and a subscriber station. Also called "subscriber loop" and "local line".

Main Distribution Frame (MDF)

A distribution frame that connects outside lines on one side and internal lines on the other.

Main Station

The subscriber's instrument (e.g., telephone or data terminal) connected to a local loop, used for originating and answering calls from the exchange.

Master

Term used to describe a point or controller which triggers an interlocking process. Example: Smoke detector designated as a master may trigger a buzzer to sound.

MAU

Media Access Unit.

MDF

Main Distribution Frame.

Mean Time Between Failure (MTBF)

The average length of time for which the system or its components work without fault.

Mean Time To Repair (MTTR)

The average time to correct a fault when the system, or its components develop one.

Measured Rate

1) A message rate structure in which the rental includes payment for a fixed number of calls within a defined area, plus a charge for additional calls. 2) A rate based on distance usage including number and length of calls.

Medium

Any substance used for the propagation of signals; in the form of modulated radio, light, or acoustic waves, from one point to another, such as optical fiber, cable, wire, dielectric slab, water, air, or free space (ISO).

Media

Physical carrier of signals (i.e, twisted pair wire, coaxial cable, fiber optics).

Megabit

One million binary bits.

Megahertz

One million Hertz (cycles per second).

Message

Sequence of characters used to convey information or data.

Message Unit (MU)

Unit of measure for charging local calls based on length of call, distance called, and time of day.

MHz

Megahertz.

MIS

Management Information System. Can be integrated into building communications network.

Mnemonic

Abbreviation for a computer instruction. Example: STA and STO are the commands for start and stop.

Modem

A contraction of modulator/demodulator. A device that translates between the digital signals used by data processing equipment and the analog signals expected by most voice switching equipment. A modem makes it possible for data to be switched over normal telephone channels and retranslated back into digital form at the receiving end.

Modulation, Amplitude (AM)

A form of modulation in which the amplitude of the carrier is varied.

Modulation, Frequency (FM)

A form of modulation in which the instantaneous frequency of a sine wave carrier is caused to depart from the carrier frequency.

MTBF

Mean Time Between Failure.

MTTA

Multi-Tenant Telecommunications Association, headquartered in Washington, DC.

MTTR

Mean Time To Repair.

MU

Message Unit.

Multidrop Line

A circuit designed to transmit data between a central site and a number of remote terminals. Terminals may transmit to the central site but not to each other. See Point-To-Point.

Multiplex

To transmit signals from two or more sources simultaneously over a single channel. Frequency division multiplexing splits a channel's total frequency band into a number of narrower bands allocated to different sources. Time division multiplexing divides the channel into successive time slots designated for certain sources so that messages from a number of devices are interleaved.

Multiplexer

A device that enables more than one signal to be sent simultaneously over one distrinct channel.

Multipoint

Type of connection where two devices are linked thru an intermediary device.

Multiprocessing

Simultaneous use of more than one processor in a multi-CPU computer system for the execution of a single user job.

Multi-Tenant Telecommunications Association (MTTA)

A trade association dedicated to the business of shared tenant services.

MUX

Multiplexer.

Nailed-Up Connection

Slang for a permanent, dedicated path through a switch.

NARUC

National Association of Regulatory Utility Commissioners.

NATA

North American Telecommunications Association.

Network Interface

The demarcation point, required by the Federal Communications Commission, between the telephone company wiring and premises distribution system wiring.

Node

1) In a topology, a point of junction of the links. 2) A switching center in data networks.

Noise

Unwanted electrical signals that degrade the performance of a communications channel.

Nonblocking

A feature of a switch where a traffic path always exists for each station. A switching environment designed not to experience a busy condition.

OA

Office automation systems. Can be integrated into building communications network.

OCC

Other Common Carrier (MCI, US Sprint, etc.)

Offline

Describes any hardware (e.g., controller, point module) which is not currently sending information.

Off-Premises Extension (OPX)

A telephone extension not located where the main switch is.

On-Hook

Telephone set with the handset resting in cradle.

On-Line

Decribes any hardware which is currently communicating.

Operating System

Organized collection of standard programs whose procedures determine a building automation system processing techniques and capabilities.

Optical Fiber

Also called a lightguide or a fiber-optic waveguide.

Optical Fiber Cable

A transmission medium consisting of a core of glass surrounded by strengthening material and a protective jacket. Signals are transmitted as light pulses, introduced into the optical fiber by a laser or light-emitting diode (LED).

Optimal Run Time

Energy-saving software program which delays the morning startup of each HVAC system until the last possible moment, yet still allows the building space to reach user-specified occupant comfort levels by occupancy time. The feature also advances the evening shutdown to the earliest possible moment which will not cause occupant discomfort prior to vacancy.

OPX

Off-Premises Extension.

OSI

Open Systems Interconnection. ISO's 7-layer reference model for network architecture used to define network protocol standards enabling all OSI-compliant computers or devices to communicate with one another.

Other Common Carrier (OCC)

Specialized common carriers (SCC), domestic and international carriers, and domestic satellite carriers engaged in providing communicative services authorized by the Federal Communications Commission.

PABX

Private Automatic Branch Exchange.

Packet

A group of binary encoded digits, including fata and call control signals, switched as a composite whole. The data, control signals, and error information are arranged in a specified format.

Packet Switching

Transmission of data by means of addressed packets.

Parallel Transmission

The simultaneous transmission of the bits making up a character or byte.

Patch Cord

A short length of cable, consisting of wires or optical fibers, used to link two cross-connect terminations. Typically, patch cords are connectorized at each end so that they can be easily unplugged for circuit rearrangements.

PBX

Private Branch Exchange; an in-house telephone system that switches calls both within a building or premises and to the telephone network.

PCM

Pulse Code Modulation.

Peripheral Device Or Equipment

A input or output unit that is not included within the primary system, e.g., printer or tape.

Pin

A conductor on a plug or connector.

Point

Each piece of equipment which is monitored or controlled by a controller, energy management system or building automation system. Examples: Fan, temperature sensor.

Point Of Presence (POP)

1) Since divestiture, the access location within a LATA of a long-distance common carrier; 2) the point at which the local telephone company terminates subscribers' circuits for long-distance, dial-up, or leased-line communications.

Point-To-Point

Describing a circuit that connects two points directly, where there are generally no intermediate processing nodes or computers, although there could be switching facilities. A type of connection, such as a phone line circuit, that links two, and only two, logical entities. See Multipoint; Broadcast.

Port

1) Point of access into a switch or network; 2) An interface through which one gains access.

POTS

Plain Old Telephone Service.

Power Fail Motor Restart

Software program which sequences the automatic restart of appropriate building equipment (e.g., large fan motors, chillers, lights) after a power failure to prevent a power surge overload.

PPM

Pulse-Position Modulation.

Primary Carrier

Long-distance carrier selected by a subscriber as the preferred provider of long-distance service.

Private Line

Denotes the channel and equipment furnished to a customer for exclusive use, usually with no access to or from the public switched telephone network. Sometimes called a leased line.

Private Network

A network established and operated by a private organization or corporation, as opposed to a public switched telephone network.

Proportional Action

A signal that is output in proportion to the amount of change in a controlled cariable (e.g., temperature sensor changing air conditioning fan speed).

Protocol

A strict, defined procedure required to initiate and maintain communications: link-by-link, end-to-end, and subscriber-to-switch.

Pseudo-Random Nature

A quality of randomness as defined by arithmetic processing.

RAM

Random Access Memory.

Rate Base

The total invested capital on which a regulated company is entitled to a reasonable rate of return.

RBHC

Regional Bell Holding Company. Sometimes also Regional Bell Operating Company (RBOC).

Realtime System

A computer system that processes information immediately on a prioritized time-basis. The building automation system is a realtime system. Example: An HVAC task (specified as a low priority) is executed after a fire task (specified as a high priority) is completed.

Regional Bell Holding Company (RBHC)

One of the seven holding companies formed at the divestiture of AT&T to provide both regulated and nonregulated telephone services. There are NYNEX, Bell Atlanta, Bell South, Ameritect, Southwestern Bell, U.S. West, Pacific Telephone.

Regulator

Device used to control flow and pressure. Example: valve.

Repeater

1) In digital transmission, equipment that receives a weary pulse train, amplifies it, retimes it, and then reconstructs the signal for retransmission. 2) In fiber optics, a device that decodes a low-power light signal, converts it electronically, and then retransmits it via an LED or laser light source.

Resale Carrier

Company that resells the services of another common carrier to the public.

Response Time

The length of time a system takes to react to a given input.

Ring Network

A network topology in which each node is connected to two adjacent nodes to form a continuous ring-like configuration.

Routing

An assignment of the path by which a message or telephone call will reach its destination.

RS-232

Serial computer communications standard.

R-232-C

EIA-specified physical interface, with associated electrical signaling, between Data Circuit terminating Equipment (DCE) and Data Terminal Equipment (DTE); they are most commonly used interface between computers and modems.

RS-422-A

An EIA specification for the electrical characteristics of balanced-voltage digital interface circuit.

RS-423-A

An EIA specification for electrical characteristics of an unbalanced-voltage digital interface circuit.

RS-449

EIA specification for general-purpose, 37-position and 9-position interface for Data Terminal Equipment (DTE) and Data Circuit Terminating Equipment (DCE) employing serial binary data interchange.

Satellite Closet

A room where cross-connect hardware is located and where cabling from information outlets is terminated.

SDLC

Synchronous Data Link Control. IBM communications line protocol providing for full-duplex transmission; used with IBM's System Network Architecture (SNA).

Service Entrance

The point at which cables enter a building.

Setpoint

Desired value assigned to a controlled variable or equipment. Examples: temperature, relative humidity, damper position.

Serial Transmission

Data transmission method whereby the bits of a character are sent sequentially on a single channel.

Shared Tenant Services (STS)

An offering that consists of any or all of the following services to tenants in a building or complex: communications, data and word processing, online data bases, HVAC control, fire and security systems, and office management.

Shielding

Prevention of electric, magnetic, or electromagnetic fields from escaping or entering an enclosed area by means of a interfering barrier. Also called (U.S. term) screening.

Shielded Pair

Two insulated wires in a cable wrapped with metallic braid or foil to prevent interference for noise-free transmission.

Signal

Aggregate waves propagated along a transmission channel intended to act on a receiving unit.

Signal-To-Noise-Ratio

Ratio of signal power to noise power.

Simplex Circuit

Circuit permitting the transmission of signals in one direction only.

Slave

A point which reacts as a response to a triggered interlock sequence. Example: If a fire alarm causes a horn to sound, the horn is the slave.

Sleeves

Short lengths of rigid metal pipes, approximately four inches in diameter, located in backbone (riser) closets, that allow cables to pass from floor to floor when closets are vertically aligned.

Slots

Openings in the floor of backbone (riser) closets that allow cables to pass through from floor to floor when closets are vertically aligned.

Smart Building

Preferred term is "Intelligent Building". Refer to Intelligent Building.

Smart Sensor

Sensors with a micro-processor packaged in the same unit to provide intelligence functions.

Softcopy

Information displayed on a CRT computer terminal.

Speakerphone

A telephone that has a speaker-microphone unit allowing hands-free conversation.

Spread Spectrum Technique

Method in which a signal is transmitted in a bandwidth greater than the frequency content of the original information.

Static

Non-changing. Any display which will not automatically change (update) even if the status of associated field points changes.

Station

One of the input or output points of a communications system - e.g., the telephone set in the telephone system. Telephone examples include rotary-dial and pushbutton telephones, key phones, speakerphones, and IVDT's.

Station Message Detail Recording (SMDR)

Records of all calls originated and/or received by a PABX system.

Subscriber Line

The telephone line connecting the exchange to the subscriber's station. Also called "access line subscriber loop".

Supervised Point

A binary point whose status is comprised of two conditions at one time: the point's contact status itself and the status of the field wiring between the point and the controller. Supervision of wiring status is typically required for UL-Listed fire and security points.

Switchhook

Switch on a telephone set, usually supporting the receiver or handset, used to signal the switching equipment or an operator during a call.

Synchronous Transmission

Transmission whereby the information and control characters are sent at regular, clocked intervals so that the sending and receiving terminals are operating in step with each other.

Tap

To monitor transmission of data through a circuit.

Tariff

A published rate for specific communications service, equipment, or facility that constitutes a contract between the user and the carrier or supplier.

TDM

Time-Division Multiplexing.

Telephone Channel

Transmission path designed for the transmission of signals representing human speech. The bandwidth allotted is usually 64K bps, but can be reduced to 32K or even 16K bps with multiplexing techniques.

Telephone Exchange

The switching center for interconnecting lines that terminate at the central office switch.

Telephone Frequency

Any frequency within the audiofrequency range essential for the transmission of speech, i.e., 300 to 3000 Hz.

Terminal

Any device capable of sending or receiving information over a communications channel.

Thermal Energy Storage (TES)

Equipment which chills water during off-peak hours (e.g., nights and weekends) for economical air conditioning of commercial buildings during periods of peak electric utility rates. Equipment can be integrated into building automation system.

Throughput

Total information processed or communicated during a specified time period.

Tie Line (TL)

A private-line communications channel.

Time-Division Multiplexing (TDM)

A means of obtaining a number of channels over a single path by dividing the path into a number of time slots and assigning each channel to a repeated time slot. At the receiving end, each time-separated channel is reassembled.

Timesharing

A sequential method of operating a computer shared by several users for different purposes at (apparently) the same time.

Ton

When used to describe chiller size it is equal to 12,000 Btu/hr of heat removal.

Topology

Physical layout of a telecommunications network (e.g., bus, ring, start).

Totally Integrated Building

Building or complex in which a single wiring plan is used for telecommunications, data processing and building automation.

Traffic

1) Messages sent and received over a communications channel. 2) A quantitative measurement of the total number of messages and their length, expressed in hundred call seconds (CCS) or other units.

Transceiver

A device that can transmit and receive traffic.

Transducer

Device that converts energy or a signal from one state to another. Example: A pneumatic to electric transducer converts a signal measured in pounds/in^2 (PSI) to a signal measured in volts (V) so it can be transmitted digitally to a building automation system.

Trend Graphics

Software program which periodically samples point information and stores this information for graphic display by the building automation system.

Trigger

The condition which causes an interlocking process to occur.

Trunk

Communication transmission paths that are used to interconnect exchanges in the main telephone network.

Turnkey System

Complete system, including hardware and software, assembled and installed by a vendor and sold as a total package.

Twisted Pair

A pair of wires formed by twisting together two single insulated wires. The twisting reduces the potential for signal interference to and from other pairs.

T1

A digital carrier facility used to transmit a DS-1 formatted signal at 1.544M bps.

Uniform Call Distribution (UCD)

This method allows calls coming in on a group of lines to be assigned stations so that all stations can handle similar loads. Call distribution systems also provide for a queuing of incoming calls with a feature first out protocol.

User Programming Language

High-level programming language which allows the operator of a building automation system to create programs. Programs can customize building automation system operations.

Vaporware

Claims for nonexistent software products (i.e., marketing hype).

VLSI

Very Large Scale Integration.

Voice Digitization

Conversion of an analog voice into digital signals for storage or transmission.

Watchtour

Software program designed to supervise and protect guards on a building tour. It monitors the progress of the guards throughout a predetermined tour using key stations. If an interruption occurs, the system signals an alarm to allow prompt backup.

WATS

Wide Area Telephone Service. Telephone company service allowing reduced costs for certain telephone call arrangements. This can be in-WATS, or 800-number service, where calls can be placed to a location from anywhere at no cost to the calling party, or out-WATS, where calls are placed out from a central location. Now available from many other common carriers.

Weekly Scheduling

The 'timeclock' software program which automatically executes operator-entered commands on a daily, time-of-day basis also determined by the operator. Example: Doors open at 8:00 MTWTHF.

Wiring Closet

A room where cables are terminated on cross-connect fields, and circuit administration can take place. There are two kinds of wiring closets: backbone (riser) closets and satellite closets.

Work Location Wiring Subsystem

The section of a premises distribution system that includes the mounting cords and connectors linking user equipment, such as telephones, data terminals, and workstations, to the information outlets in rooms and offices.

X.25

CCITT recommendation that specifies the interface between user Data Terminal Equipment (DTE) and Packet-Switching Data Circuit-terminating Equipment (DCE).

XM

Cross Modulation Distortion.

INDEX